Studies in Fuzziness and Soft Computing   307

Editor-in-Chief

Prof. Janusz Kacprzyk
Systems Research Institute
Polish Academy of Sciences
ul. Newelska 6
01-447 Warsaw
Poland
E-mail: kacprzyk@ibspan.waw.pl

For further volumes:
http://www.springer.com/series/2941

Eulalia Szmidt

Distances and Similarities in Intuitionistic Fuzzy Sets

 Springer

Eulalia Szmidt
Systems Research Institute
Polish Academy of Sciences
Warsaw
Poland

ISSN 1434-9922   ISSN 1860-0808 (electronic)
ISBN 978-3-319-03302-0   ISBN 978-3-319-01640-5 (eBook)
DOI 10.1007/978-3-319-01640-5
Springer Cham Heidelberg New York Dordrecht London

Printed on acid-free paper

Springer is part of Springer Science+Business Media (www.springer.com)

Foreword

The book I am glad to write my foreword to is a very relevant position in literature on intuitionistic fuzzy sets or, maybe even more generally, in the fuzzy set theory.

In virtually all application issues related to the very essence of similarity, distances are crucial. Just to quote some more important examples, let me mention data analysis and data mining, machine learning, decision theory and analysis, control etc. Of course, this short list is by no means exhaustive.

Professor Eulalia Szmidt presents in this excellent book a brief, yet an extremely informative and constructive account of various tools and techniques to effectively and efficiently define and determine similarity in the case of objects described in terms of intuitionistic fuzzy sets.

The problem of quantifying similarity and distance can be viewed from at least two perspectives. On the one hand it is a conceptual level, and Professor Szmidt provides here a survey of both well known and original new concepts and measures. On the other hand, the calculations of similarity, dissimilarity, and distance in non trivial cases have an explicit algorithmic character, and many computational problems may appear. Here, again, Professor Szmidt provides the reader with a full-fledged set of procedures that can be used to solve practical problems.

To summarize, we should congratulate the Author for writing an excellent book, which certainly constitutes a part of the literature that will be widely used by the scientific community.

Sofia, June 2013   *Krassimir Atanassov*

Contents

Chapter 1
Introduction

Dealing with imprecise information is a common task and challenge in everyday life, as uncertainty is inevitably involved in every real world system. Models are constructed to control, predict, and diagnose such systems, and so uncertainty should be properly incorporated into system description.

For a long time dealing with uncertain information was a big challenge. Until the 1960s, uncertainty was considered solely in terms of probability theory and understood as randomness. This seemingly unambiguous connection between uncertainty and probability was paralleled by several mathematical theories, distinct from probability theory, which are able to characterize situations under uncertainty.

The best known of these theories, which began to emerge in 1960s, are: the theory of fuzzy sets (Zadeh [254]), evidence theory (Dempster [57], [58], [59], Shafer [150]), possibility theory (Zadeh [256]), the theory of fuzzy measures (Sugeno [155], [156]), rough set theory (Pawlak [132]), and lately – theory of intuitionistic fuzzy sets (Atanassov [4], [15], [22]) which stimulates an increasing number of researchers all over the world.

Development of the theories mentioned demonstrated that there are several distinct types of uncertainty (Klir [104], [106], Klir and Yuan [109], Klir and Wierman [108]), Klir and Folger [107]) and as the term *uncertainty* was for three ages connected with randomness, in further considerations we will mainly use a more general term – *imperfect information* (which includes also randomness).

To view intuitionistic fuzzy sets in a proper perspective and as a tool for representing imperfect information, it may be expedient to look first at the emergence and significance of the fuzzy set theory. It was proposed by Zadeh in 1965 as a simple and efficient tool for representing and processing imprecise concepts and quantities, exemplified by, say, *tall men, large numbers* etc. which fall beyond the scope of conventional precise mathematics, because their very essence is a gradual (not abrupt) transition between the membership and non-membership of elements in a set. Zadeh attained this goal by replacing the conventional characteristic function of the classical "crisp" set, which takes on its values in {0,1} by the so-called membership function, which takes on the values in the interval [0,1], allowing for the representation of membership to a degree.

E. Szmidt, *Distances and Similarities in Intuitionistic Fuzzy Sets,*
Studies in Fuzziness and Soft Computing 307,
DOI: 10.1007/978-3-319-01640-5_1, © Springer International Publishing Switzerland 2014

In other words, it is necessary to specify for each element of a fuzzy set a real value from the unit interval, this value denoting the membership degree of the given element. Clearly, one can view this requirement to be too strict because for an imprecise concept we need to specify a precise number. Such a reasoning has led Goguen [74], [75] to propose the so-called L-fuzzy sets in which the set of values of the membership function is an ordered set. So, the L-fuzzy sets lower the requirements on the available information as in L-fuzzy sets one does not need to know the precise values of the degrees of membership.

The concept of an intuitionistic fuzzy set can be viewed in this perspective as another approach to the definition of a fuzzy set in the case when available information is not sufficient for the definition of imprecise concept by means of a fuzzy set as outlined above. The basic assumption of the fuzzy set theory is, namely, that if we specify the degree of membership of an element in a fuzzy set as a real number from [0,1], say a, then the degree of its non-membership is automatically determined as $1 - a$. This assumption need not hold. Instead, in the intuitionistic fuzzy set theory it is assumed that the value of non-membership should only be at most equal $1 - a$ (in other words, the sum of membership and non-membership is at most 1). The difference between the sum of membership and non-membership and the value of 1 lets us express the lack of knowledge (the hesitancy concerning both membership and non-membership of an element to a set). In this way we can better model imprecise information. It is worth emphasizing again that analysis and modeling of any real world phenomenon or process must take into account various, inherent facets of imperfect information.

The inevitable connection between imperfect information and decision making is explained by Shackle [151]:

> In a predestinate world, decision would be illusory; in a world of a perfect foreknowledge, *empty*; in a world without natural order, *powerless*. Our intuitive attitude to life implies non-illusory, non-empty, non-powerless decision .... Since decision in this sense excludes both perfect foresight and anarchy in nature, it must be defined as choice in face of bounded uncertainty.

To deal successfully with imperfect information one needs to have a proper model to represent the information, and adequate measures making it possible to process this information. Among such measures the concepts of distance and similarity measures play the leading role.

One could expect that from the fact that the intuitionistic fuzzy sets are a generalization of the fuzzy sets, the measures of distances, and similarity between the intuitionistic fuzzy sets are just straightforward generalizations of their respective counterparts for the fuzzy sets. We examine carefully this idea, verifying two types of distances and similarity measures, being the consequences of two types of representations of the intuitionistic fuzzy sets.

The first representation of the intuitionistic fuzzy sets takes into account the membership values and the non-membership values only, and indeed, it is a straightforward generalization of the fuzzy sets (additionally taking into account in an explicit manner the non-membership values). We call it two term representation of the intuitionistic fuzzy sets, and consider the corresponding measures of the distances and similarity measures, which make use of the two terms only.

The second type of representation of the intuitionistic fuzzy sets that we use, takes into account the membership values, the non-membership values, and the hesitation margin values, hence three terms, and so we call it three term representation. This type of representation generates respective distances, and similarity measures (which make use of the three terms).

Both types of representations are correct from the mathematical point of view. The same concerns their corresponding measures. However, from the point of view of decision making the results we obtain for the respective measures in the cases of the two types of representations differ. We examine the problem in details.

Another issue we deal with is the well known correspondence (duality) of the concepts of distance and similarity for crisp sets and for fuzzy sets. We consider whether just the same interrelations are valid for the intuitionistic fuzzy sets.

We consider here only discrete models.

We are deeply convinced of the usefulness of the intuitionistic fuzzy sets in widely understood decision making, just because of their natural ability to express imprecise information (by taking – explicitly – into account the hesitation margins).

This opinion is supported in a sense by the general rule one should have in mind while constructing a model (the key tool in decision making), i.e., by attempt to maximize its usefulness. Some helpful hints how to do it are given by Klir and Wierman [108]:

> The aim is closely connected with the relationship among three key characteristics of every systems model: *complexity, credibility,* and *imprecision*. This relationship, which is a subject of current study in systems science, is not as yet fully understood. We only know that uncertainty has a pivotal role in any efforts to maximize the usefulness of systems models. Although usually undesirable when considered alone, imprecision becomes very valuable when considered in connection to the other characteristics of systems models: a slight increase in imprecision may often significantly reduce complexity and, at the same time, increase credibility of the model. Imprecision is thus an important commodity in the modeling business, a commodity which can be treated for gains in the other essential characteristic of models.

We need only adequate (efficient) tools to express imprecision. The intuitionistic fuzzy sets seem to be such a tool.

The purpose of this book is first of all to present the state of the art in the area of quantifying of similarity and distances in the context of the intuitionistic fuzzy sets. However, it also contains new elements regarding the measures of similarity and distance, both in terms of new definitions and computational algorithms.

The scope of this book is as follows.

In Chapter 2, first, we briefly present crisp sets, fuzzy sets, and intuitionistic fuzzy sets. Basic concepts, operations, and relations are given, so as to show that the intuitionistic fuzzy sets are a generalization of the fuzzy sets.

Next, we consider two types of geometrical representations of the intuitionistic fuzzy sets - taking into account two terms (membership and non-membership values), i.e., the 2D representation, and taking into account all three terms (membership and non-membership values, as well as hesitation margins) characterizing the intuitionistic fuzzy sets, i.e., the 3D representation.

Next, we discuss interrelationships among crisp sets, fuzzy sets, and intuitionistic fuzzy sets.

Chapter 2 terminates with two chosen methods of deriving the intuitionistic fuzzy sets from data. The automatic method provided there may be useful from the point of view of applications.

Chapter 3 is devoted to distances between the intuitionistic fuzzy sets. Distances are discussed for two intuitionistic fuzzy set representations, namely the two term representation (only membership and non-membership values are taken into account), and the three term representation (membership values, non-membership values and hesitation margins are accounted for). Both representations are mathematically correct.

After the basic definitions have been introduced, the norms and metrics over the intuitionistic fuzzy sets in the two term representation are presented.

Next, the distances between intuitionistic fuzzy sets in the three term representation are discussed. The geometrical and analytical arguments are presented to indicate why from the decision making point of view the three term representation is important. Special attention is devoted to Hausdorff distances – it is shown that the method working well for the interval-valued fuzzy sets does not work properly for the intuitionistic fuzzy sets. We also show some problems with the Hausdorff distance when the Hamming metric is applied for the case of the two term intuitionistic fuzzy set representation.

Chapter 3 ends with a development of application of the distances introduced in a measure for ranking the intuitionistic fuzzy alternatives. Again, the three term approach for calculating the distances (i.e. taking into account membership values, non-membership values, and the hesitation margin values) is shown to be justified in terms of usefulness.

In Chapter 4 similarity measures between the intuitionistic fuzzy sets are discussed. Axiomatic relations between fuzzy similarity measures and fuzzy distance measures are presented. However, it is shown that distance alone does not suffice to conclude about similarity between the intuitionistic fuzzy sets.

Some similarity measures, proposed in the respective literature, which make use of the two term representation of the intuitionistic fuzzy sets, are recalled and their reliability is discussed. Next, another array of similarity measures, making use of the three term representation of the intuitionistic fuzzy sets and taking into account the complements of the elements compared, is presented.

Chapter 4 closes with an extended analysis of the Pearson-like correlation coefficient between the intuitionistic fuzzy sets. The coefficient discussed, like Pearson's coefficient between crisp sets, measures how strong is relationship between the intuitionistic fuzzy sets, and indicates if the sets are positively or negatively correlated. Again, all three terms characterizing the intuitionistic fuzzy sets are taken into account, and importance of each of them is stressed in a series of thorough numerical tests on two well known data benchmarks from the UC, Irvine, the Pima Indian diabetes and Iris data.

Finally, conclusions and bibliography are provided.

We use in this book the term "intuitionistic fuzzy sets", although there have been vivid discussion concerning this term in 2005. However, since that time, Professor Atanassov has published many papers with new operators (especially, but not uniquely, negations) which support his point of view as far as the name is concerned. So we do not repeat the old arguments in the new situation (new publications) and do not wish to start a new discussion about the term. We refer the interested reader to the new book by Professor Atanassov where the subject is addressed: Atanassov K.T. On Intuitionistic Fuzzy Sets Theory. Springer-Verlag 2012.

Chapter 2
Intuitionistic Fuzzy Sets as a Generalization of Fuzzy Sets

Abstract. In the mid-1980s Atanassov introduced the concept of an intuitionistic fuzzy set. Basically, his idea was that unlike the conventional fuzzy sets in which imprecision is just modeled by the membership degree from [0,1], and for which the non-membership degree is just automatically the complementation to 1 of the membership degree, in an intuitionistic fuzzy set both the membership and non-membership degrees are numbers from [0,1], but their sum is not necessarily 1. Thus, one can express a well known psychological fact that a human being who expresses the degree of membership of an element in a fuzzy set, very often does not express, when asked, the degree of non-membership as the complementation to 1. This idea has led to an interesting theory whose point of departure is such a concept of intuitionistic fuzzy set. In this chapter we give brief introduction to intuitionistic fuzzy sets. After recalling main definitions, concepts, operations and relations over crisp sets, fuzzy sets, and intuitionistic fuzzy sets we discuss interrelationships among the three types of sets. Two geometrical representations of the intuitionistic fuzzy sets, useful in further considerations are discussed. Finally, two approaches of constructing the intuitionistic fuzzy sets from data are presented. First approach is via asking experts. Second one – the automatic, and mathematically justified method to construct the intuitionistic fuzzy sets from data seems to be especially important in the context of analyzing information in big data bases.

2.1 Main Definitions

As Atanassov's concept of an intuitionistic fuzzy set can be viewed as a generalization of a fuzzy set definition in the case when available information is not sufficient for the definition of an imprecise concept by means of a conventional fuzzy set, we start from the basic definitions of a set, and of a fuzzy set.

2.1.1 Crisp Sets (Classical Sets)

A set is one of the basic concepts in mathematics. Informally, a set is a collection of objects (elements) having similar properties (attributes). A classical set A (crisp set) has sharp boundaries, i.e., there are two possibilities only:

E. Szmidt, *Distances and Similarities in Intuitionistic Fuzzy Sets,*
Studies in Fuzziness and Soft Computing 307,
DOI: 10.1007/978-3-319-01640-5_2, © Springer International Publishing Switzerland 2014

– an element x belongs to the set $(x \in A)$, or
– an element does not belong to the set $(x \notin A)$.

A classical set can be expressed by its characteristic function.

Definition 2.1. For a set X and a subset A of X $(A \subseteq X)$ we call

$$\varphi_A(x) = \begin{cases} 1 & \text{if } x \in A \\ 0 & \text{if } x \notin A \end{cases} \tag{2.1}$$

the characteristic function of the set A in X.

Using the notion of the characteristic function $\varphi_A(x)$, a crisp set A can be given as :

$$A = \{<x, \varphi_A(x) > /x \in X\} \tag{2.2}$$

For a conventional (crisp) set an element can not belong "to some extent" to a set.

2.1.2 Fuzzy Sets

A generalization of a crisp set is a fuzzy set. The notion of a fuzzy set was introduced by Zadeh [254]. A fuzzy set A' in a universe of discourse X is characterized by a membership function $\mu_{A'}$ which assigns to each element $x \in X$ a real number $\mu_{A'}(x) \in [0,1]$ expressing the membership grade of x in the fuzzy set A'.

Definition 2.2. (Zadeh [254])
A fuzzy set A' in $X = \{x\}$ is given by (Zadeh [254]):

$$A' = \{<x, \mu_{A'}(x) > /x \in X\} \tag{2.3}$$

where $\mu_{A'} : X \to [0,1]$ is the membership function of the fuzzy set A'; $\mu_{A'}$ for every element $x \in X$ describes its extent of membership to fuzzy set A'.

As mentioned above, a fuzzy set A' is a generalization of a conventional (crisp) set A represented by its characteristic function $\varphi_A : X \to \{0,1\}$ (2.1). Full membership of x in A' occurs for $\mu'_A(x) = 1$, full non-membership is for $\mu'_A(x) = 0$ but opposite to a classical set other membership degrees are also allowed.

Every crisp set is a fuzzy set. The membership function of a crisp set $A \subseteq X$ can be expressed as its characteristic function

$$\mu_A(x) = \begin{cases} 1 & \text{if } x \in A \\ 0 & \text{if } x \notin A \end{cases} \tag{2.4}$$

2.1.3 Intuitionistic Fuzzy Sets

The notion of an intuitionistic fuzzy set was introduced by Atanassov (Atanassov [4]). An intuitionistic fuzzy set is a generalization of a fuzzy set.

Definition 2.3. (Atanassov [4], [6], [15] [22])
An intuitionistic fuzzy set A in X is given by:

$$A = \{<x, \mu_A(x), \nu_A(x) > /x \in X\} \tag{2.5}$$

where

$$\mu_A : X \rightarrow [0,1]$$
$$\nu_A : X \rightarrow [0,1]$$

with the condition

$$0 \leq \mu_A(x) + \nu_A(x) \leq 1 \quad \forall x \in X$$

The numbers $\mu_A(x)$ and $\nu_A(x)$ denote, respectively, the degrees of membership and non-membership of the element $x \in X$ to the set A.

Obviously, every fuzzy set corresponds to the following intuitionistic fuzzy set:

$$FS : \{<x, \mu_A(x), 1 - \mu_A(x) > /x \in X\}. \tag{2.6}$$

Definition 2.4. (Atanassov [4], [6], [15] [22])
For an intuitionistic fuzzy set A we will call

$$\pi_A(x) = 1 - \mu_A(x) - \nu_A(x) \tag{2.7}$$

the **intuitionistic fuzzy index** (hesitation margin) of the element x in the set A. The $\pi_A(x)$ expresses the lack of knowledge on whether x belongs to A or not.

It is obvious that

$$0 \leq \pi_A(x) \leq 1 \quad \text{for every } x.$$

For every fuzzy set A', where $x \in X$

$$\pi_A(x) = 1 - \mu_{A'}(x) - [1 - \mu_{A'}(x)] = 0.$$

The hesitation margin turns out to be important while considering the distances (Szmidt and Kacprzyk [165], [171], [188], entropy (Szmidt and Kacprzyk [175], [192]), similarity (Szmidt and Kacprzyk [193]) for the intuitionistic fuzzy sets, i.e., the measures that play crucial role in virtually all information processing tasks. The hesitation margin is shown to be indispensable also in the ranking of intuitionistic fuzzy alternatives as it indicates how reliable (sure) information presented for an alternative is (cf. Szmidt and Kacprzyk [198], [205]).

Making use of the intuitionistic fuzzy sets instead of fuzzy sets implies the introduction of additional degrees of freedom (non-memberships and hesitation margins) into the set description. Such a generalization of fuzzy sets gives us an additional possibility to represent imprecise knowledge which may lead to describing many real problems in a more adequate way.

It is woth stressing that from the point of view of the applications, taking into account the hesitation margins (besides non-membership values) is crucial in many areas exemplified by image processing (cf. Bustince et al. [45], [46]), classification of imbalanced and overlapping classes (cf. Szmidt and Kukier [229], [230], [231]), group decision making, negotiations, voting and other situations (cf. Szmidt and Kacprzyk [164], [167], [172], [174], [177], [178], [189]).

Intuitionistic fuzzy sets based models may be adequate mainly in the situations when we face human testimonies, opinions, etc. involving answers of three types:

– yes,
– no,
– abstaining i.e. which can not be classified (because of different reasons, eg. "I do not know", "I am not sure", "I do not want to answer", "I am not satisfied with any of the options" etc.).

Below we present an example given by Atanassov (Atanassov [15]). The example illustrates the essence of the intuitionistic fuzzy sets, and stresses the differences between them and the fuzzy sets.

Example 2.1. (Atanassov [15]) Let X be the set of all countries with elective governments. Assume that we know for every country $x \in X$ the percentage of the electorate who have voted for the corresponding government. Let it be denoted by $M(x)$ and let $\mu(x) = \frac{M(x)}{100}$. Let $v(x) = 1 - \mu(x)$. This number corresponds to that part of electorate who have not voted for the government. By means of the fuzzy set theory we cannot consider this value in more detail. However, if we define $v(x)$ as the number of votes given to parties or persons outside the government, then we can show the part of electorate who have not voted at all and the corresponding number will be $1 - \mu(x) - v(x)$. Thus, we can construct the set $\{\langle x, \mu(x), v(x)\rangle | x \in X\}$ and obviously, $0 \le \mu(x) + v(x) \le 1$. □

2.2 Brief Introduction to Fuzzy Sets

2.2.1 Basic Concepts

A fuzzy set A' in X is said to be empty, $A' = \emptyset$, if and only if

$$\mu'_A(x) = 0, \qquad \text{for each } x \in X \tag{2.8}$$

Two fuzzy sets A' and B' in the same universe of discourse X are said to be equal, i.e., $A' = B'$, if and only if

$$\mu'_A(x) = \mu'_B(x), \qquad \text{for each } x \in X \tag{2.9}$$

A fuzzy set A' defined in X is a subset of a fuzzy set B' in X, $A' \subseteq B'$, if and only if

$$\mu'_A(x) \le \mu'_B(x), \qquad \text{for each } x \in X \tag{2.10}$$

A fuzzy set A' defined in X is said to be normal if and only if the membership function takes on the value of 1 for at least one value of its argument, i.e.

$$\max_{x \in X} \mu'_A(x) = 1 \qquad (2.11)$$

Otherwise, a fuzzy set is said to be subnormal.

The support $\mathrm{supp}A'$ of a fuzzy set A' in X is the following (nonfuzzy, i.e. crisp) set:

$$\mathrm{supp}A' = \{x \in X : \mu'_A(x) > 0\} \qquad (2.12)$$

where $\emptyset \subseteq \mathrm{supp}A' \subseteq X$.

The α-cut (or α-level set) A_α, of a fuzzy set A' in X is defined as the following (nonfuzzy) set:

$$A_\alpha = \{x \in X : \mu_A(x) \geq \alpha\}, \qquad \text{for each } \alpha \in (0,1] \qquad (2.13)$$

and if "$\geq$" in (2.13) is replaced by "$>$," then we have the strong α-cut (or strong α-level set), of a fuzzy set A' in X.

The α-cuts are quite important from the point of view of both theory and applications as they make it possible to uniquely replace a fuzzy set by a sequence of nonfuzzy sets. More details and properties of α-cuts can be found in any book on fuzzy set theory (cf. e.g., Dubois and Prade [61], Klir and Folger [107], or Klir and Yuan [109]).

Another important issue is to define how many elements are contained in a fuzzy set, i.e. to define its cardinality. The most often used definition is given below.

A nonfuzzy cardinality of a fuzzy set $A' = \mu'_A(x_1)/x_1 + \cdots + \mu'_A(x_n)/x_n$, the so-called *sigma-count* $\Sigma\mathrm{Count}(A')$, is defined as (Zadeh [256], [257])

$$\Sigma\mathrm{Count}(A') = \sum_{i=1}^{n} \mu'_A(x_i) \qquad (2.14)$$

Other definitions of cardinality making use of α-cuts were proposed by Zadeh [257] (see also Kacprzyk [97]). More discussion, criticism, and new definitions of cardinality are given by Ralescu [139], and Wygralak [247].

2.2.2   Selected Operations on Fuzzy Sets

Just like in the case of crisp sets, basic operations on fuzzy sets are complement, union, and intersection. They are defined in terms of the respective membership functions. The operations presented below correspond to the operations on intuitionistic fuzzy sets (cf. Section 2.3.2).

The complement A'^C of a fuzzy set A' in X, corresponds to negation "not", and is defined as

$$\mu_{A'^c}(x) = 1 - \mu'_A(x), \qquad \text{for each } x \in X \qquad (2.15)$$

The intersection of two fuzzy sets A' and B' in X, written $A' \cap B'$, corresponds to the connective "and", and is defined as

$$\mu_{A' \cap B'}(x) = \mu'_A(x) \wedge \mu'_B(x), \qquad \text{for each } x \in X \qquad (2.16)$$

where "$\wedge$" is the minimum operation, i.e. $a \wedge b = \min(a,b)$.

The union of two fuzzy sets A' and B' in X, written $A' + B'$, corresponds to the connective "or", and is defined as

$$\mu_{A' + B'}(x) = \mu'_A(x) \vee \mu'_B(x), \qquad \text{for each } x \in X \qquad (2.17)$$

where "$\vee$" is the maximum operation, i.e. $a \vee b = \max(a,b)$.

More general than the intersection and the union defined above, are so-called t-norms and s-norms (t-conorms).

A t-norm is defined as

$$t : [0,1] \times [0,1] \longrightarrow [0,1] \qquad (2.18)$$

such that, for each $a,b,c \in [0,1]$ the following properties are fulfilled:

1. the value 1 is the unit element, i.e.

$$t(a,1) = a$$

2. monotonicity, i.e.

$$a \le b \Longrightarrow t(a,c) \le t(b,c)$$

3. commutativity, i.e.

$$t(a,b) = t(b,a)$$

4. associativity, i.e.

$$t[a,t(b,c)] = t[t(a,b),c]$$

A t-norm is monotone non-decreasing in both arguments, and $t(a,0) = 0$. Among the most used t-norms there are:

- the minimum

$$t(a,b) = a \wedge b = \min(a,b) \qquad (2.19)$$

  which is the most widely used,
- the algebraic product

$$t(a,b) = a \cdot b \qquad (2.20)$$

- Łukasiewicz t-norm

$$t(a,b) = \max(0, a+b-1) \qquad (2.21)$$

An s-norm called also t-conorm is defined as

$$s : [0,1,] \times [0,1] \longrightarrow [0,1] \tag{2.22}$$

such that, for each $a,b,c \in [0,1]$ the following properties are fulfilled:

1. the value 0 is the unit element, i.e.

$$s(a,0) = a$$

2. monotonicity

$$a \leq b \Longrightarrow s(a,c) \leq s(b,c)$$

3. commutativity, i.e.

$$s(a,b) = s(b,a)$$

4. associativity, i.e.

$$s[a,s(b,c)] = s[s(a,b),c]$$

The most used s-norms include:

- the maximum

$$s(a,b) = a \vee b = \max(a,b) \tag{2.23}$$

which is the most widely used,
- the probabilistic product

$$s(a,b) = a + b - ab \tag{2.24}$$

- Łukasiewicz s-norm

$$s(a,b) = \min(a+b,1) \tag{2.25}$$

It is worth noticing that a t-norm is dual to an s-norm in that

$$s(a,b) = 1 - t(1-a,1-b) \tag{2.26}$$

Another concept, crucial from the point of view of theory and application, is that of a relation. It is discussed both for fuzzy sets and intuitionistic fuzzy sets in Section 2.3.3.

2.3   Brief Introduction to the Intuitionistic Fuzzy Sets

We will start this section by discussing two geometrical representations of intuitionistic fuzzy sets. The representation using two terms (membership values and non-membership values) describing the intuitionistic fuzzy sets leads to the so called 2D representation. Using three terms (membership values, non-membership values, and hesitation margins) in the intuitionistic fuzzy set description results in the so called 3D representation.

Other geometrical representations of the intuitionistic fuzzy sets are given by Atanassov [12], [13], [14], [15], [22]).

Later on, basic operators and relations over the intuitionistic fuzzy sets will be discussed.

2.3.1 Two Geometrical Representations of the Intuitionistic Fuzzy Sets

Having in mind that for each element x belonging to an intuitionistic fuzzy set A, the values of membership, non-membership and the intuitionistic fuzzy index sum up to one, i.e.

$$\mu_A(x) + \nu_A(x) + \pi_A(x) = 1 \tag{2.27}$$

and that each of the membership, non-membership, and the intuitionistic fuzzy index belongs to $[0,1]$, we can imagine a unit cube (Figure 2.1) inside which there is an *MNH* triangle where the above equation is fulfilled (Szmidt and Kacprzyk [171]).

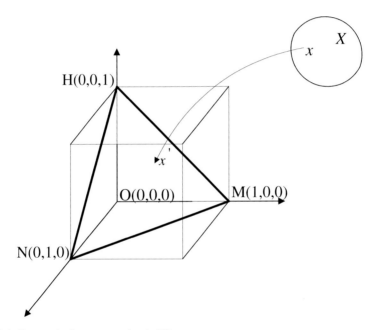

Fig. 2.1 Geometrical representation in 3D

In other words, the *MNH* triangle represents a surface where coordinates of any element belonging to an intuitionistic fuzzy set can be represented. Each point belonging to the *MNH* triangle is described via three coordinates: (μ, ν, π). Points M and N represent crisp elements. Point $M(1,0,0)$ represents elements fully belonging to an intuitionistic fuzzy set as $\mu = 1$. Point $N(0,1,0)$ represents elements fully not belonging to an intuitionistic fuzzy set as $\nu = 1$. Point $H(0,0,1)$ represents elements about which we are not able at all to say if they belong or not to an intuitionistic fuzzy set (intuitionistic fuzzy index $\pi = 1$). Such an interpretation is intuitively appealing and provides means for the representation of many aspects of imperfect information. The segment *MN* (where $\pi = 0$) represents elements belonging to classical fuzzy sets ($\mu + \nu = 1$).

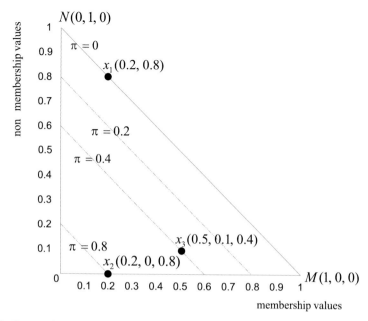

Fig. 2.2 Geometrical representation in 2D

Any other combination of the values characterizing an intuitionistic fuzzy set can be represented inside the triangle *MNH*. In other words, each element belonging to an intuitionistic fuzzy set can be represented as a point (μ, v, π) belonging to the triangle *MNH* (Figure 2.1).

It is worth mentioning that the geometrical interpretation is directly related to the definition of an intuitionistic fuzzy set introduced by Atanassov [4], [15], and it does not need any additional assumptions.

Remark: We use the capital letters (e.g., *M*, *N*, *H*) for the geometrical representation of x_i's (Figure 2.2) on the plane. The same notation (capital letters) is used in this book for sets, but we always explain the current meaning of a symbol used.

Another possible geometrical representation of an intuitionistic fuzzy set can be in two dimensions (2D) – Figure 2.2 (cf. Atanassov [15]). It is worth noticing that although we use a 2D figure (which is more convenient to draw in many cases), we still adopt our approach (e.g., Szmidt and Kacprzyk [171], [188], [175], [192], [193], [218]) taking into account all three terms (membership, non-membership and hesitation margin values) describing the intuitionistic fuzzy sets. As previously, any element belonging to an intuitionistic fuzzy set may be represented inside an *MNO* triangle (*O* is projection of *H* in Figure 2.1). Each point belonging to the *MNO* triangle is still described by the three coordinates: (μ, v, π), and points *M* and *N* represent, as previously, crisp elements. Point $M(1,0,0)$ represents elements fully belonging to an intuitionistic fuzzy set as $\mu = 1$, and point $N(0,1,0)$ represents elements fully not belonging to an intuitionistic fuzzy set as $v = 1$. Point $O(0,0,1)$

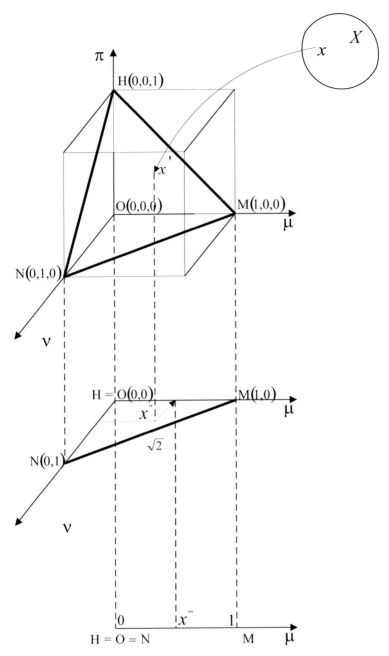

Fig. 2.3 Illustration of the interrelations between 3D and 2D representations of intuitionistic fuzzy sets

represents elements about which we cannot say if they belong or not to an intuitionistic fuzzy set (the intuitionistic fuzzy index $\pi = 1$). Segment MN (where $\pi = 0$) represents elements belonging to the classic fuzzy sets ($\mu + v = 1$). For example, point $x_1(0.2, 0.8, 0)$ (Figure 2.2), like any element of the segment MN, represents an element of a fuzzy set. A line parallel to MN describes the elements with the same values of the hesitation margin. In Figure 2.2 we can see point $x_3(0.5, 0.1, 0.4)$ representing an element with the hesitation margin equal 0.4, and point $x_2(0.2, 0, 0.8)$ representing an element with the hesitation margin equal 0.8. The closer a line that is parallel to MN is to O, the higher the hesitation margin.

In Figure 2.3 (Szmidt [158]) relations between the 2D and 3D representations are presented. It is worth stressing that 2D representation of the intuitionistic fuzzy sets (Figure 2.2), i.e., the triangle MNO is the orthogonal projection of the triangle MNH (3D representation – Figure 2.1) on the plane (Figure 2.3, the upper and the middle parts). The orthogonal projection transfers $x' \in MNH$ into $x'' \in MNO$. Segment MN represents a fuzzy set described by μ and v.

The orthogonal projection of the segment MN on the axis μ (the segment $[0, 1]$ is only considered) gives the fuzzy set represented by μ only (Figure 2.3, its bottom part). This orthogonal projection transfers $x'' \in MNO$ into $x''' \in OM$.

2.3.2   Operations Over the Intuitionistic Fuzzy Sets

This section contains results introduced by Atanassov [15] on operations and relations over the intuitionistic fuzzy sets. The point of departure is constituted by the respective definitions of relations and operations over fuzzy sets which are extended here. In the reverse perspective, relations and operations on fuzzy sets turn out to be particular cases of these new definitions.

Here is the definition of basic relations and operations on intuitionistic fuzzy sets.

Definition 2.5. (Atanassov [15])
For every two intuitionistic fuzzy sets A and B the following relations and operations can be defined ("iff" means "if and only if"):

$$A = B \text{ iff } (\forall x \in X)(\mu_A(x) = \mu_B(x) \& v_A(x) = v_B(x)), \tag{2.28}$$

$$A^C = \{\langle x, v_A(x), \mu_A(x)\rangle | x \in X\}, \tag{2.29}$$

$$A \cap B = \{\langle x, \min(\mu_A(x), \mu_B(x)), \max(v_A(x), v_B(x))\rangle | x \in X\}, \tag{2.30}$$

$$A \cup B = \{\langle x, \max(\mu_A(x), \mu_B(x)), \min(v_A(x), v_B(x))\rangle | x \in X\}, \tag{2.31}$$

$$\begin{aligned} A + B = \{\langle x, \mu_A(x) \\ + \mu_B(x) - \mu_A(x).\mu_B(x), v_A(x).v_B(x)\rangle, | x \in X\} \end{aligned} \tag{2.32}$$

$$A.B = \{\langle x, \mu_A(x).\mu_B(x), \nu_A(x) \\ +\nu_B(x) - \nu_A(x).\nu_B(x)\rangle \mid x \in X\}, \tag{2.33}$$

$$A@B = \{\langle x, (\frac{\mu_A(x)+\mu_B(x))}{2}, \frac{(\nu_A(x)+\nu_B(x))}{2}\rangle \mid x \in X\}, \tag{2.34}$$

$$A\$B = \{\langle x, \sqrt{\mu_A(x).\mu_B(x)}, \sqrt{\nu_A(x).\nu_B(x)}\rangle \mid x \in X\}, \tag{2.35}$$

$$A*B = \{\langle x, \frac{\mu_A(x)+\mu_B(x)}{2.(\mu_A(x).\mu_B(x)+1)}, \\ \frac{\nu_A(x)+\nu_B(x)}{2.(\nu_A(x).\nu_B(x)+1)}\rangle \mid x \in X\}, \tag{2.36}$$

$$A\bowtie B = \{\langle x, 2.\frac{\mu_A(x).\mu_B(x)}{\mu_A(x)+\mu_B(x)}, 2.\frac{\nu_A(x).\nu_B(x)}{(\nu_A(x)+\nu_B(x))}\rangle \mid x \in X\}, \tag{2.37}$$

for which we will accept that if

$$\mu_A(x) = \mu_B(x) = 0, \text{ then } \frac{\mu_A(x).\mu_B(x)}{\mu_A(x)+\mu_B(x)} = 0$$

$$\text{and if } \nu_A(x) = \nu_B(x) = 0, \text{ then } \frac{\nu_A(x).\nu_B(x)}{\nu_A(x)+\nu_B(x)} = 0.$$

Operations (2.28)–(2.37) have also their counterparts in the fuzzy set theory.

Example 2.2. (Atanassov [15])
Let $X = \{a,b,c,d,e\}$, let the intuitionistic fuzzy sets A and B have the form $\{< x_i, \mu(x_i), \nu(x_i) >\}$:

$$A = \{\langle a,0.5,0.3\rangle, \langle b,0.1,0.7\rangle, \langle c,1.0,0.0\rangle, \langle d,0.0,0.0\rangle, \langle e,0.0,1.0\rangle\},$$

$$B = \{\langle a,0.7,0.1\rangle, \langle b,0.3,0.2\rangle, \langle c,0.5,0.5\rangle, \langle d,0.2,0.2\rangle, \langle e,1.0,0.0\rangle\}.$$

Then

$$\overline{A} = \{\langle a,0.3,0.5\rangle, \langle b,0.7,0.1\rangle, \langle c,0.0,1.0\rangle, \langle d,0.0,0.0\rangle, \langle e,1.0,0.0\rangle\},$$

$$A\cap B = \{\langle a,0.5,0.3\rangle, \langle b,0.1,0.7\rangle, \langle c,0.5,0.5\rangle, \langle d,0.0,0.2\rangle, \langle e,0.0,1.0\rangle\},$$

$$A\cup B = \{\langle a,0.7,0.1\rangle, \langle b,0.3,0.2\rangle, \langle c,1.0,0.0\rangle, \langle d,0.2,0.0\rangle, \langle e,1.0,0.0\rangle\},$$

$$A+B = \{\langle a,0.85,0.03\rangle, \langle b,0.37,0.14\rangle, \langle c,1.0,0.0\rangle, \langle d,0.2,0.0\rangle, \\ \langle e,1.0,0.0\rangle\},$$

$$A.B \quad = \{\langle a,0.35,0.37\rangle, \langle b,0.03,0.76\rangle, \langle c,0.5,0.5\rangle, \langle d,0.0,0.2\rangle,$$
$$\langle e,0.0,1.0\rangle\},$$

$$A@B \ = \{\langle a,0.6,0.2\rangle, \langle b,0.2,0.45\rangle, \langle c,0.75,0.25\rangle, \langle d,0.1,0.1\rangle,$$
$$\langle e,0.5,0.5\rangle\},$$

$$A\$B \ = \{\langle a,0.591...,0.173...\rangle, \langle b,0.173...,0.374...\rangle, \langle c,0.0707...,0.0\rangle,$$
$$\langle d,0.0,0.0\rangle, \langle e,0.0,0.0\rangle\},$$

$$A*B \ = \{\langle a,0.444...,0.194...\rangle, \langle b,0.194...,0.394...\rangle, \langle c,0.5,0.5\rangle,$$
$$\langle d,0.1,0.1\rangle, \langle e,0.5,0.5\rangle\},$$

$$A\bowtie B = \{\langle a,0.583...,0.15\rangle, \langle b,0.15,0.311...\rangle, \langle c,0.666...,0.0\rangle,$$
$$\langle d,0.0,0.0\rangle, \langle e,0.0,0.0\rangle\}.$$

$\square$

Proposition 2.1. (Atanassov [15])
For every three intuitionistic fuzzy sets A, B and C, following properties hold: :

$$A\cap B = B\cap A, \tag{2.38}$$

$$A\cup B = B\cup A, \tag{2.39}$$

$$A + B = B + A, \tag{2.40}$$

$$A.B = B.A, \tag{2.41}$$

$$A@B = B@A, \tag{2.42}$$

$$A\$B = B\$A, \tag{2.43}$$

$$A\bowtie B = B\bowtie A, \tag{2.44}$$

$$A*B = B*A, \tag{2.45}$$

$$A\cap B)\cap C = A\cap(B\cap C), \tag{2.46}$$

$$(A\cup B)\cup C = A\cup(B\cup C), \tag{2.47}$$

$$(A + B) + C = A + (B + C), \tag{2.48}$$

$$(A.B).C = A.(B.C), \tag{2.49}$$

$$(A \cap B) \cup C = (A \cup C) \cap (B \cup C), \tag{2.50}$$

$$(A \cap B) + C = (A + C) \cap (B + C), \tag{2.51}$$

$$(A \cap B).C = (A.C) \cap (B.C), \tag{2.52}$$

$$(A \cap B)@C = (A@C) \cap (B@C), \tag{2.53}$$

$$(A \cap B) \bowtie C = (A \bowtie C) \cap (B \bowtie C), \tag{2.54}$$

$$(A \cup B) \cap C = (A \cap C) \cup (B \cap C), \tag{2.55}$$

$$(A \cup B) + C = (A + C) \cup (B + C), \tag{2.56}$$

$$(A \cup B).C = (A.C) \cup (B.C), \tag{2.57}$$

$$(A \cup B)@C = (A@C) \cup (B@C), \tag{2.58}$$

$$(A \cup B) \bowtie C = (A \bowtie C) \cup (B \bowtie C), \tag{2.59}$$

$$(A + B).C \subset (A.C) + (B.C), \tag{2.60}$$

$$(A + B)@C \subset (A@C) + (B@C), \tag{2.61}$$

$$(A.B) + C \supset (A + C).(B + C), \tag{2.62}$$

$$(A.B)@C \supset (A@C).(B@C), \tag{2.63}$$

$$(A@B) + C = (A + C)@(B + C), \tag{2.64}$$

$$(A@B).C = (A.C)@(B.C), \tag{2.65}$$

$$A \cap A = A, \tag{2.66}$$

$$A \cup A = A, \tag{2.67}$$

$$A @ A = A, \tag{2.68}$$

$$A \$ A = A, \tag{2.69}$$

$$A \bowtie A = A, \tag{2.70}$$

$$\overline{\overline{A} \cap \overline{B}} = A \cup B, \tag{2.71}$$

$$\overline{\overline{A} \cup \overline{B}} = A \cap B, \tag{2.72}$$

$$\overline{\overline{A} + \overline{B}} = A.B, \tag{2.73}$$

$$\overline{\overline{A}.\overline{B}} = A + B, \tag{2.74}$$

$$\overline{\overline{A} @ \overline{B}} = A @ B, \tag{2.75}$$

$$\overline{\overline{A} \$ \overline{B}} = A \$ B, \tag{2.76}$$

$$\overline{\overline{A} \bowtie \overline{B}} = A \bowtie B, \tag{2.77}$$

$$\overline{\overline{A} * \overline{B}} = A * B. \tag{2.78}$$

2.3.2.1 The "necessity" and "possibility" Operators

The operators over intuitionistic fuzzy sets, presented above, correspond to the respective operators over fuzzy sets. Here we present two operators introduced by Atanassov in 1983, which are "meaningless" (Atanassov [4], [15]) in the case of fuzzy sets.

Definition 2.6. (Atanassov [4], [15]) Let us define, for every intuitionistic fuzzy set A, the following operators:

- the necessity operator

$$\Box A = \{\langle x, \mu_A(x), 1 - \mu_A(x)\rangle | x \in X\}, \tag{2.79}$$

- the possibility operator

$$\Diamond A = \{\langle x, 1 - v_A(x), v_A(x)\rangle | x \in X\}. \tag{2.80}$$

If A is a classical fuzzy set, then

$$\Box A = A = \Diamond A. \tag{2.81}$$

From (2.81) it follows that both "$\Box$" (2.79) and "$\Diamond$" (2.80) are meaningless for a fuzzy set. Atanassov [15] considers in length the properties, modifications, and extensions of "$\Box$" (2.79) and "$\Diamond$" (2.80). Here we only recall their two extensions.
Let $\alpha \in [0,1]$ be a fixed number.

Definition 2.7. (Atanassov [15]) Given an intuitionistic fuzzy set A, an operator D_α is defined as follows:

$$D_\alpha(A) = \{\langle x, \mu_A(x) + \alpha.\pi_A(x), v_A(x) + (1 - \alpha).\pi_A(x)\rangle | x \in X\} \tag{2.82}$$

where $\alpha \in [0,1]$.
From definition 2.7 we see that $D_\alpha(A)$ is a fuzzy set, namely:

$$\mu_A(x) + \alpha.\pi_A(x) + v_A(x) + (1 - \alpha).\pi_A(x) = \mu_A(x) + v_A(x) + \pi_A(x) = 1.$$

Several interesting properties of $D_\alpha(A)$ (2.82) are given by proposition 2.2:

Proposition 2.2. (Atanassov [15]) For every intuitionistic fuzzy set A and for every $\alpha, \beta \in [0,1]$:

$$\text{if } \alpha \leq \beta, \text{ then } D_\alpha(A) \subset D_\beta(A), \tag{2.83}$$

$$D_0(A) = \Box A, \tag{2.84}$$

$$D_1(A) = \Diamond A. \tag{2.85}$$

The operator D_α is a generalization of the operators "necessity" and "possibility". The operator D_α has been be extended even further. Namely, Atanassov [15] introduced operator $F_{\alpha,\beta}$ (2.86).
Let $\alpha, \beta \in [0,1]$ and $\alpha + \beta \leq 1$.

Definition 2.8. (Atanassov [7]) The operator $F_{\alpha,\beta}$, for an intuitionistic fuzzy set A, is defined as:

$$F_{\alpha,\beta}(A) = \{\langle x, \mu_A(x) + \alpha.\pi_A(x), v_A(x) + \beta.\pi_A(x)\rangle | x \in X\}. \tag{2.86}$$

The above operators are not only important from the theoretical point of view (indicating that the intuitionistic fuzzy sets are a generalization of the fuzzy sets) but they are also important from the point of view of applications. The operator $D_\alpha(A)$ was applied in constructing a classifier recognizing imbalanced classes (Szmidt and Kukier [229], [230], [231]). The operator $F_{\alpha,\beta}$ has been applied for image recognition (Bustince et al. [131], [46], [48]).

2.3.3   *Intuitionistic Fuzzy Relations*

As it was shown (cf. definitions in Section 2.1), one term, i.e., membership function, fully describes a fuzzy set, whereas two terms are necessary when we discuss the intuitionistic fuzzy sets. Similar differences hold when we define a fuzzy relation and an intuitionistic fuzzy relation.

Definition 2.9. A fuzzy relation between two non-fuzzy sets $X = \{x\}$ and $Y = \{y\}$ is defined on a Cartesian product $X \times Y$, i.e. $R \subset X \times Y = \{(x,y) : x \in X, y \in Y\}$, and given by

$$R = \{< (x,y), \mu_R(x,y) > /x \in X, y \in Y\} \tag{2.87}$$

where $\mu_R : X \times Y \to [0,1]$ is a membership function of a fuzzy relation R assigning to every pair (x,y), $x \in X$, $y \in Y$, its degree of membership $\mu_R(x,y) \in [0,1]$ describing the measure of intensity of a fuzzy relation R between x and y.

Example 2.3. If $X = \{Al, Bob, Clark\}$ and $Y = \{Paul, Jim\}$, the fuzzy relation R labelled "resemblance" may be exemplified by

$$R = (Al, Paul)/0.5 + (Al, Jim)/0.3 + (Bob, Paul)/0.6 +$$
$$+ (Bob, Jim)/0.4 + (Clark, Paul)/0.9 + (Clark, Jim)/0.1$$

to be read as: there is resemblance between Paul and Clark (with respect to "our own" subjective aspects) to degree 0.9, i.e. to a very high extent, and rather low resemblance between Jim and Clark - to degree 0.1 only, etc.

Any fuzzy relation (in a finite $X \times Y$) may be represented in a matrix form. The following matrix corresponds to the above relation "resemblance"

$$R = [r_{ij}] = \begin{array}{c|cc} & Paul & Jim \\ \hline Al & 0.5 & 0.3 \\ Bob & 0.6 & 0.4 \\ Clark & 0.9 & 0.1 \end{array}$$

□

Taking into account the definition of intuitionistic fuzzy set, the definition of an intuitionistic fuzzy relation R can be introduced as a counterpart of fuzzy relation.

Definition 2.10. (Atanassov [5], [23], [15])
An intuitionistc fuzzy relation between two non-fuzzy sets $X = \{x\}$ and $Y = \{y\}$ is defined on a Cartesian product $X \times Y$, i.e. $R \subset X \times Y = \{(x,y) : x \in X, y \in Y\}$, and given by

$$R = \{< (x,y), \mu_R(x,y), \nu_R(x,y) > /x \in X, y \in Y\} \tag{2.88}$$

with the condition

$$0 \le \mu_A(x) + \nu_A(x) \le 1 \quad \forall x \in X, y \in Y$$

where μ_R - as before, $\nu_R : X \times Y \rightarrow [0,1]$ is a non-membership function of a fuzzy relation R assigning to every pair (x,y), $x \in X$, $y \in Y$, its degree of non-membership $\nu_R(x,y) \in [0,1]$, being the measure of falsity of a fuzzy relation R between x and y.

Therefore, an intuitionistic fuzzy relation is described by any two terms from the triplet: membership function, non-membership function, intuitionistic fuzzy index function.

Example 2.4. If $X = \{Bob, Peter\}$ and $Y = \{Liz, Jim, John\}$, the intuitionistic fuzzy relation R labelled "cooperation" while preparing a new project, may be given as $\{(x,y), \mu_R(x,y), \pi_R(x,y)\}$, i.e. intuitionistic fuzzy relation can be also described by giving $\pi_R(x,y)$ instead of $\nu_R(x,y)$

$$R = (Bob, Liz)/1,0 + (Bob, Jim)/0.7,0.2 + (Bob, John)/0.5,0.3 +$$
$$+ (Peter, Liz)/0.7,0.2 + (Peter, Jim)/0.9,0 + (Peter, John)/0.4,0.5$$

or, in a matrix form

$$\mu_R = \begin{array}{c|cc} & Bob & Peter \\ \hline Liz & 1 & 0.7 \\ Jim & 0.7 & 0.9 \\ John & 0.5 & 0.2 \end{array} \qquad \pi_R = \begin{array}{c|cc} & Bob & Peter \\ \hline Liz & 0 & 0.2 \\ Jim & 0.2 & 0 \\ John & 0.3 & 0.8 \end{array}$$

to be read as: excellent cooperation between Bob and Liz is foreseen (to the highest degree: 1), and any difficulties are not expected ($\pi = 0$) about their cooperation. Peter and Jim usually cooperate to a very high extent (to degree 0.9) but in a very rare situations it is known that they have quite different opinions ($\pi = 0$ which means $\nu = 0.1$). Peter and John usually at the beginning do not agree ($\mu = 0.2$ only), but they are open for arguments ($\pi = 0.8$) so it is possible in almost all cases to convince them to go for the same goal, etc.   □

2.4 Interrelationships: Crisp Sets, Fuzzy Sets, Intuitionistic Fuzzy Sets

On the basis of the definitions and properties presented in this chapter, the following conclusions can be drawn:

- A membership function fully describes a fuzzy set (by specifying a membership function we automatically know the non-membership function).
- If we want to describe fully an intuitionistic fuzzy set, we must use any two terms from the triplet: {*membership function, non-membership function, intuitionistic fuzzy index function*}.

In other words, applying intuitionistic fuzzy sets instead of fuzzy sets means introducing another degree of freedom into the set description (apart from a function μ_A there appears a function ν_A or π_A).

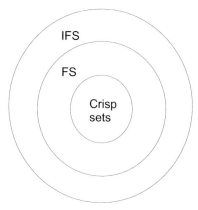

Fig. 2.4 An illustration of interrelations among conventional sets, fuzzy sets (FS) and intuitionistic fuzzy sets (IFS)

Interrelations among conventional sets, fuzzy sets (FS) and intuitionistic fuzzy sets (IFS) are given in Figure 2.4.

The meaning of Figure 2.4 is explained in Figure 2.5. For a crisp set, two points only in the coordinate system given in Figure 2.5 can represent the elements belonging to such a set. As an element corresponding to a crisp set fully belongs or fully does not belong to a crisp set, only the point with coordinates: $\mu = 1$ and $v = 0$ (fully belonging), or another point: $\mu = 0$ and $v = 1$ (fully not-belonging) can represent the elements from crisp sets. The case is illustrated in the upper part of Figure 2.5.

In the case of a fuzzy set, because of the fact that $\mu + v = 1$, besides previously described points (fully belonging, and fully not-belonging), also the entire segment connecting these points can be an image of elements belonging to a fuzzy set. The middle part of Figure 2.5 illustrates this fact.

Finally, for an intuitionistic fuzzy set, because of the condition: $0 \leq \mu + v \leq 1$, not only the points described above (the segment with its ends), but also the interior of the shaded triangle at the bottom part of Figure 2.5 can represent the elements belonging to an intuitionistic fuzzy set. Every one of the parallel lines (inside the triangle) is an image of the elements with the same value of intuitionistic fuzzy index.

In the above sense, the intuitionistic fuzzy sets contain fuzzy sets which, in turn, contain conventional (crisp) sets (cf. Figure 2.4). In terms of information it means that in the case of

- crisp sets – information is complete (elements fully belong or fully do not belong to a crisp set),
- fuzzy sets – information is also complete but the elements can belong to a fuzzy set to some degree; knowing the degree to which the elements belong to a fuzzy set (μ), we immediately know the degree to which they do not belong to the set ($1 - \mu$),

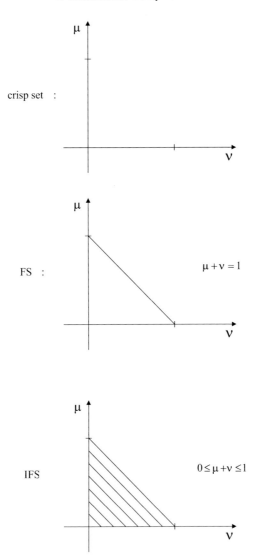

Fig. 2.5 Comparison of the possible images (in the same coordinate system) of elements belonging to crisp sets, fuzzy sets (FS) and intuitionistic fuzzy sets (IFS)

- intuitionistic fuzzy sets – information may be not complete because of the intuitionistic fuzzy indices; on the other hand, using the intuitionistic fuzzy sets we may express the same information as via crisp sets or fuzzy sets (cf. Figures 2.4 and 2.5).

In the next section, starting from relative frequency distributions, we describe the automatic algorithm of deriving the intuitionistic fuzzy sets from data (Szmidt and Baldwin [162]).

2.5 Deriving the Intuitionistic Fuzzy Sets from Data

In order to apply the intuitionistic fuzzy sets, which seem to be a very good tool for representation and processing of imperfect information, one should be able to construct their membership and non-membership functions. In this section we propose two ways of deriving the membership and non-membership functions for the intuitionistic fuzzy sets:

a) by asking experts;
b) from relative frequency distributions (histograms).

The second (automatic) method is justified by some similarities/parallels between intuitionistic fuzzy set theory and mass assignment theory – a well known tool for dealing with both probabilistic and fuzzy uncertainties. We also recall a semantic for membership functions – the interpretation having its roots in the possibility theory. Both mass assignment theory and a semantic interpretation of the membership functions made it possible to construct the automatic algorithm assigning the functions describing the intuitionistic fuzzy sets.

Uncertainty was identified and expressed for several centuries (starting from the mid-seventeenth century) in terms of probability theory only. Uncertainty was a synonym of randomness. This situation changed in the 1960s when other theories, distinct from probability theory, characterizing different aspects of uncertain situations, were introduced. Uncertainty started to be perceived as a multidimensional concept for which randomness became one of its dimension only. Other components turned out equally important from the point of view of representing and processing information.

The theory of fuzzy sets (Zadeh [254]), theory of evidence (Dempster-Shafer theory [59, 150]), possibility theory (Zadeh [256]), theory of fuzzy measures (Sugeno [155]) became the most visible theories dealing with different aspects of uncertainty. Here we explore two of the theories dealing with widely understood uncertainty, namely, intuitionistic fuzzy set theory (Atanassov [4]) which is a generalization of fuzzy set theory, and mass assignment theory (Baldwin [30, 29]) related to the theory of evidence (but the constraint that $m(0) = 0$ is not imposed). The theory of mass assignment we apply also differs from the Dempster-Shafer theory of evidence since the method of combining mass assignments is different. The theory discussed here is consistent with probability theory.

We start from showing some similarities/parallels between intuitionistic fuzzy set theory [Atanassov [6, 15]] and mass assignment theory (Baldwin [30], Baldwin et al. [38, 35]). The similarities we stress do not mean that one of the theories could replace the other or is better. To the contrary, the similarities we show seem to be important from the point of view of further development of both theories. Here, the noticed similarities made it possible to construct an algorithm of the automatic derivation of the membership and non-membership functions for intuitionistic fuzzy sets.

Making use of positive and (independently given) negative information, which is the core of the intuitionistic fuzzy set approach, is natural in real life, and as an obvious consequence, is well-known in psychology [e.g., Sutherland [157], Kahneman [98]]. The idea also attracted attention of the scientists in soft computing. It would be difficult to deal with machine learning (making use of examples and counter-examples), modeling of preferences or voting without taking into account positive and (independent) negative data.

Atanassov and Gargov [24]) noticed in 1989 that from the mathematical point of view intuitionistic fuzzy sets are equipollent to interval-valued fuzzy sets. However, from the point of view of solving problems (starting from the stage of collecting data), both kind of sets are different. The intuitionistic fuzzy sets make a user consider independently positive and negative information whereas when employing interval-valued fuzzy sets, user's attention is focused on positive information (in an interval) only. This fact, strongly connected with a psychological phenomenon called by the Nobel Prize winner Kahneman (cf. Kahneman [98]) "bounded rationality" (see also Sutherland [157]), caused among others by the fact that people tend to notice and take into account only most obvious aspects (e.g. advantages only), places the intuitionistic fuzzy sets among the up-to-date means of knowledge representation and processing.

To apply the intuitionistic fuzzy sets one should be able to assign respective membership values and non-membership values. Here we discuss two ways of assigning the membership and non-membership values for the intuitionistic fuzzy sets: by asking experts, and from the relative frequency distributions (histograms).

The models applying the intuitionistic fuzzy sets may be especially useful in the situations when we face human testimonies, opinions, etc. involving answers of three types:

– yes,
– no,
– abstaining, i.e. such that can not be classified in the former two (because of different reasons, eg. "I do not know", "I am not sure", "I do not want to answer", "I am not satisfied with any of the options", etc.).

Example 2.5. (Szmidt and Baldwin [162]) Let us assume that each individual x_i from a set X of n individuals who vote for/against building of nuclear power plant (electors voting for/against a given candidate or his opponent, judges voting for/against acquittal, consumers expressing/not expressing interest in buying a product) belongs to

• a set of individuals (judges, electors) voting for – to the extent $\mu(x_i)$,
• a set of individuals voting against – to the extent $v(x_i)$.

It is worth emphasizing that by means of the fuzzy set theory it is not possible to consider the situation in more details. By means of intuitionistic fuzzy set theory we can also point out

• a set of individuals who did not answer neither "yes" nor "no"– to the extent $\pi(x_i)$,
  whereas: $\mu_A(x) + v_A(x) + \pi_A(x) = 1$; $\pi(x_i)$ is an intuitionistic fuzzy index.

From the point of view of e.g. market analysts (election committees) it seems rather interesting to be able to assess the above data in terms of the possible final results of voting, by giving intervals containing

- probability of voting for

$$Pr_{for} \in [\mu, \mu + \pi]$$

where:

$$\mu = \frac{1}{n} \sum_{i=1}^{n} \mu(x_i)$$

$$\pi = \frac{1}{n} \sum_{i=1}^{n} \pi(x_i)$$

- probability of voting against

$$Pr_{against} \in [\nu, \nu + \pi]$$

where:

$$\nu = \frac{1}{n} \sum_{i=1}^{n} \nu(x_i)$$

with the condition $Pr_{for} + Pr_{against} = 1$.

Interpreting the above results in terms of mass assignment (see Section 2.5.2) we could say that the necessary support for is equal to μ, the necessary support against is equal to ν, whereas the possible support for (the best possible result) Pos^+ is equal to $\mu + \pi$, and the possible support against (the worst possible result) Pos^- is equal to $\nu + \pi$.

It is necessary to stress that we have made a simplifying assumption in the above example by assigning a sign of equality to probabilities and memberships/non-memberships. This assumption is valid under the condition that each value of membership/non-membership occurs with the same probability for each x_i. Here, for the sake of simpler notation, we follow this assumption. However, in general, probabilities for the intuitionistic fuzzy sets are calculated as discussed in (Szmidt and Kacprzyk [169, 170]) and recalled by Definitions 2.11 and 2.12.

Definition 2.11. (Szmidt and Kacprzyk [169, 170]) By an intuitionistic fuzzy event A we will mean an intuitionistic fuzzy subset belonging to the elementary event space X, i.e. $A \subset X$ whose membership function $\mu_A(x)$, non-membership function $\nu_A(x)$, and intuitionistic fuzzy index $\pi_A(x)$ are Borel measurable.

Definition 2.12. (Szmidt and Kacprzyk [169, 170]) Let us assign to every element of an intuitionistic fuzzy event $A \subset X = \{x_1, ..., x_n\}$ (where X is the elementary event space) its probability of occurrence, i.e. $p(x_1), ..., p(x_n)$.
Minimal probability $p_{\min}(A)$ of an intuitionistic fuzzy event A is equal to

$$p_{\min}(A) = \sum_{i=1}^{n} p(x_i)\mu(x_i)$$

Maximal probability of an intuitionistic fuzzy event A is equal to

$$p_{\max}(A) = p_{min}(A) + \sum_{i=1}^{n} p(x_i)\pi(x_i)$$

so probability of an event A is a number from the interval $[p_{min}(A), p_{max}(A)]$, or

$$p(A) \in [\sum_{i=1}^{n} p_A(x_i)\mu_A(x_i), \sum_{i=1}^{n} p_A(x_i)\mu_A(x_i) +$$
$$+ \sum_{i=1}^{n} p_A(x_i)\pi_A(x_i)], \tag{2.89}$$

probability of a complement event A^C is a number from the interval $[p_{min}(A^C), p_{max}(A^C)]$, or

$$p(A^C) \in [\sum_{i=1}^{n} p_A(x_i)\nu_A(x_i), \sum_{i=1}^{n} p_A(x_i)\nu_A(x_i) +$$
$$+ \sum_{i=1}^{n} p_A(x_i)\pi_A(x_i)] \tag{2.90}$$

Applications of the intuitionistic fuzzy sets to group decision making, negotiations and other real situations are presented, e.g., in (Szmidt and Kacprzyk [163, 164, 167, 173, 177, 178, 179, 182]).

The question arises how to derive the membership and non-membership functions.

2.5.1 Derivation of the Intuitionistic Fuzzy Sets by Experts

We will discuss the problem of deriving membership and non-membership functions for the intuitionistic fuzzy models in the simplest case – when one person considers one decision only (this simple case can be easily extended to more complicated situations - with more persons and more decisions).

Assume that somebody considers a problem of changing his/her job. To decide if a new job is interesting enough to give up a previous one it seems reasonable to prepare a whole list of questions. The list would depend on the personal preferences but in general the following questions presented in Table 2.1 seem to be important (Szmidt and Baldwin [162].

Assuming that all the questions are equally important in Table 2.1, we can immediately conclude how to evaluate the considered case – just by summing up:

- all the positive answers (7/12) - this is the value of the membership for the considered option,
- all the negative answers (3/12) - this is the value of the non-membership for the considered option,

Table 2.1 The questions considered when changing a job

No	Questions	+/?/-
1	Is the job interesting	+
2	Salaries	−
3	Possibilities of promotion	?
4	Expected pension	−
5	Number of hours spent in work	?
6	Holidays – how long	+
7	Is the work safe	+
8	Responsibility	+
9	Time of the travel: home–work	−
10	Social reputation	+
11	Necessary creativity	+
12	Connected stress	+

- all the answers for which it was impossible to say "yes" or "not" (2/12) - this is the value of the intuitionistic fuzzy index for the considered option.

We can notice that employing the intuitionistic fuzzy sets just forces an individual to consider both advantages (membership values) and disadvantages (non-membership values) of a considered solution. Next, the imprecise area is taken into account as well. The importance of such an approach lies in the fact that most people concentrate usually on one or two "most visible" aspects of a problem. They do not try to find out the contrary arguments or to consider uncertain (in wide sense, i.e. not restricted to randomness) aspects of a situation (cf. Sutherland [157]).

The structure of the intuitionistic fuzzy sets make us consider a situation/problem taking into account more aspects. We refer again an interested reader to (Szmidt and Kacprzyk [163, 164, 167, 173, 177, 178, 179, 182]) where we exploit this fact - using the intuitionistic fuzzy sets to group decision making. In short, the problem boils down to selecting an option or a set of options which are best accepted by most of the individuals. The options are considered in pairs. Employing the intuitionistic fuzzy sets forces each individual to look at each pair (i,j) of the options considering: advantages of the first option over the second one (membership value), disadvantages of the first option over the second one (non-membership value), and taking into account lack of knowledge (intuitionistic fuzzy index) as far as the two options are concerned. In other words, the intuitionistic fuzzy sets force a user to explore a problem from different points of view – including all important aspects which should be taken into account but, unfortunately, are often omitted by people making decisions. This fact, strongly connected with a phenomenon called by the Nobel Prize winner Kahneman (cf. Kahneman [98]) "bounded rationality", caused, in particular, by the framing effect (explained in terms of salience and anchoring, playing a central role in treatment of judgements and choice), makes the intuitionistic fuzzy sets a highly effective means of knowledge representation and processing.

2.5.2 Automatic Method of Deriving Intuitionistic Fuzzy Sets from Relative Frequency Distributions (Histograms)

Baldwin (Baldwin [30], Baldwin et al. [38, 35]) developed the theory of mass assignment to provide a formal framework for manipulating both probabilistic and fuzzy uncertainty.

A fuzzy set can be converted into a mass assignment (Baldwin [28], Dubois and Prade [62]). This mass assignment represents a family of probability distributions.

Definition 2.13. (Mass Assignment) (Baldwin et al. [29], [35], [32], [37]) Let A' be a fuzzy subset of a finite universe X such that the range of the membership function of A', is $\{\mu_1,...,\mu_n\}$ where $\mu_i > \mu_{i+1}$. Then the mass assignment of A' denoted $m_{A'}$, is a probability distribution on 2^X satisfying

$$m_{A'}(F_i) = \mu_i - \mu_{i+1} \text{ where } F_i = \{x \in X | \mu(x) \geq \mu_i\}$$
$$\text{for } i = 1,...,n \tag{2.91}$$

We call the sets $F_1,...,F_n$ the focal elements of $m_{A'}$. The details of mass assignment theory are presented by Baldwin et al. [38].

Example 2.6. (Baldwin [31])
For $X = \{x_1, x_2, x_3, x_4\}$,
if $A' = x_1/1 + x_2/0.7 + x_3/0.4 + x_4/0.3$
then the associated mass assignment is
$m_{A'} = x_1 : 0.3, \quad \{x_1, x_2\} = 0.3, \quad \{x_1, x_2, x_3\} = 0.1, \quad \{x_1, x_2, x_3, x_4\} = 0.3$

The basic representation of uncertainty in the language FRIL [Baldwin et al. [38, 32]) are the so called Support Pairs which are associated with mass assignments and represent intervals containing unknown probabilities. Support Pairs are used to characterize uncertainty in facts and conditional probabilities in rules. A Support Pair (n,p) comprises a necessary and possible support and can be identified with an interval in which the unknown probability lies. Baldwin and Pilsworth [33] gave a voting interpretation of a support pair – the lower (necessary) support n represents the proportions of a sample population voting in favor of a proposition, whereas $(1-p)$ represents the proportion voting against; $(p-n)$ represents the proportion abstaining.

On the other hand, considering a voting model in terms of the intuitionistic fuzzy sets, [cf. Example 2.1 (Atanassov [15])] we have

- the membership values μ are equal to the proportion of a sample population voting in favour of a proposition,
- the non-membership values v are equal to the proportion of a sample population voting against,
- the values of the hesitation margin π represents the proportion abstaining.

The interpretation of the parameters from Baldwin's voting model, and from intuitionistic fuzzy set (abbreviated IFS) voting model is presented in Table 2.2.

Table 2.2 Equality of the parameters for Baldwin's voting model and IFS voting model

	Baldwin's voting model	IFS voting model
voting in favour	n	μ
voting against	$1 - p$	ν
abstaining	$p - n$	π

In other words, we can represent a Support Pair (n, p) using notation of the intuitionistic fuzzy sets by the following simple expression (Szmidt and Baldwin [162], [159], [160]):

$$(n, p) = (n, n + p - n) = (\mu, \mu + \pi) \tag{2.92}$$

i.e.: using notation of the intuitionistic fuzzy sets a Support Pair from the Baldwin's voting model can be expressed.

Moreover, one can note that the necessary support for the statement not being true is equal to one minus the possibility of the support for the statement being true, i.e. $1 - p$. Similarly, the possible support for the statement being not true is one minus the necessary support for the statement being true i.e. $1 - n$. Having in mind the correspondence of these parameters, we can express this fact making use of the intuitionistic fuzzy set notation, namely

$$(1 - p, 1 - n) = (\nu, \nu + \pi)$$

The following three Support Pairs (n, p) are especially interesting (Baldwin and Pilsworth [33]):

- $(1, 1)$ expresses total support for the associated statement,
- $(0, 0)$ characterizes total support against and
- $(0, 1)$ represents complete uncertainty in the support.

Certainly, the meaning of the above Support Pairs is just the same in the models expressed in terms of the intuitionistic fuzzy sets (assuming that we consider probabilities for intuitionistic fuzzy membership/non-membership values as it was explained in the context of Definition 2.12):

- $(1, 1)$ means that $\mu = 1$ and $\pi = 0$, i.e. total support,
- $(0, 0)$ means $\mu = 0$ and $\pi = 0$ which involves $\nu = 1$, i.e. total support against,
- $(0, 1)$ means $\mu = 0$ and $\pi = 1$ i.e.: complete lack of knowledge concerning support.

So, to sum up, both the Support Pairs and the intuitionistic fuzzy set models give the same intervals containing the probability of the fact being true. The difference between the upper and lower bounds of the intervals is a measure of the uncertainty associated with the fact [160], [159].

As Baldwin [34] has observed, the mass assignment structure is best used to represent knowledge that is statistically based in the sense that the values can be measured, even if the measurements themselves are approximate or uncertain.

Next notion, useful in our considerations, is so called least prejudiced distribution [62], [38].

For A', a fuzzy subset of a finite universe X, the least prejudiced distribution of A', denoted $lp_{A'}$, is a probability distribution on X given by

$$lp_{A'}(x) = \sum_{F_i:x\in F_i} \frac{m_{A'}(F_i)}{|F_i|} \qquad (2.93)$$

A mass assignment corresponding to a normalized fuzzy subset of X naturally generates a family of probability distributions on X where each distribution corresponds to some redistribution of the masses associated with sets to elements of those sets. The most intuitive seems to redistribute a priori the mass associated with a set in the uniform manner among the elements of that set. In effect we obtain the distribution which coincides with Smet's pignistic probability [153], and with Dubois and Prade's [62] possibility-probability transformation based on a generalized Laplacean indifference principle.

It is worth emphasizing that the least prejudiced distribution (2.93) provides a mechanism by which we can, in a sense, convert a fuzzy set into a probability distribution. That is, in the absence of any prior knowledge, we might on being told A' naturally infer the distribution $lp_{A'}$ relative to a uniform prior. Certainly, if fuzzy sets are to serve as descriptions of probability distributions, the converse must also hold. The least prejudiced distribution provides the bijective possibility-probability transformation. In other words, for a probability distribution P on a finite universe X there is a unique fuzzy set A' conditioning on which yields this distribution (Baldwin et al. [37], Dubois and Prade [63]).

The mass assignment theory has been applied in some fields, such as induction of decision trees [36], computing with words among others, giving good results for real data.

Theorem 2.1. (*Baldwin et al.* [37]) *Let P be a probability distribution on a finite universe X taking the range of values $\{p_1,...,p_n\}$ where $0{\leq}p_{i+1} < p_i{\leq}1$ and $\sum_{i=1}^{n} p_i = 1$. Then P is the least prejudiced distribution of a fuzzy set A' if and only if A' has a mass assignment given by*

$$m_{A'}(F_i) = \mu_i - \mu_{i+1} \quad for\ i = 1,...,n-1$$
$$m_{A'}(F_n) = \mu_n$$

where

$$F_i = \{x \in X | P(x) \geq p_i\}$$
$$\mu_i = |F_i|p_i + \sum_{j=i+1}^{n} (|F_j| - |F_{j+1}|)p_j \qquad (2.94)$$

The proof is given in (Baldwin at al. [37]).

Dubois and Prade [63] proposed a bijection method identical with the above algorithm, but it is worth mentioning that the motivation in [37] is quite different. A similar approach to mapping between probability and possibility was considered by Yager [248]. Yamada [251] has given a further justification for the transformation.

To sum up, Theorem 2.1 provides a general procedure converting a relative frequency distribution into a fuzzy set, i.e. gives us means for generating fuzzy sets from data. As the membership values of a fuzzy set univocally assign the non-membership values, Theorem 2.1 fully describes a fuzzy set.

Moreover, Theorem 2.1 gives an idea how to convert the relative frequency distributions into an intuitionistic fuzzy set. However, when discussing intuitionistic fuzzy sets we consider membership values and independent non-membership values [cf. (2.5)–(2.6)]. In result, Theorem 2.1 gives only a partial description we look for. To obtain a complete description of an intuitionistic fuzzy set (with independent membership and non-membership values), the procedure as in Theorem 2.1 should be carried out twice. Consequently, we obtain two fuzzy sets. To interpret the two fuzzy sets properly in terms of the intuitionistic fuzzy sets we recall first a semantic for membership functions.

Depending on the particular applications, Dubois and Prade [64] have explored three main semantics for membership functions. Here we make use of the interpretation proposed by Zadeh [256] when he introduced the possibility theory. Membership $\mu(x)$ is there the degree of possibility that a parameter x has the value μ (Zadeh [256]).

In effect of repeating the procedure as in Theorem 2.1 two times (first – for data representing membership values, second – for data representing non-membership values), and taking into account the interpretation that the obtained values are the degrees of possibility, we obtain the following results (Szmidt and Baldwin [162]).

- First time the algorithm from Theorem 2.1 is performed for the relative frequencies connected with membership values. In effect we obtain (fuzzy) possibilities $Pos^+(x) = \mu(x) + \pi(x)$ that x has the value Pos^+.
  $Pos^+(x)$ (left hand side of the above equation) means the values of a membership function for a fuzzy set (possibilities). In terms of intuitionistic fuzzy sets (right hand side of the above equation) these possibilities are equal to possible (maximal) memberships of an intuitionistic fuzzy set, i.e.
  $\mu(x) + \pi(x)$, where $\mu(x)$ – the values of the membership function for an intuitionistic fuzzy set, and $\mu(x) \in [\mu(x), \mu(x) + \pi(x)]$.
- Second time the algorithm from Theorem 2.1 is performed for the (independent) relative frequencies connected with non-membership values. In effect we obtain (fuzzy) possibilities $Pos^-(x) = \nu(x) + \pi(x)$ that x has not the value Pos^-.
  $Pos^-(x)$ (left hand side of the above equation) means the values of a membership function for another (than in the previous step) fuzzy set (possibilities). In terms of the intuitionistic fuzzy sets (right hand side of the above equation) these possibilities are equal to the possible (maximal) non-membership values, i.e.
  $\nu(x) + \pi(x)$, where $\nu(x)$ – the values of the non-membership function for an intuitionistic fuzzy set, and $\nu(x) \in [\nu(x), \nu(x) + \pi(x)]$.

The Algorithm of Constructing the Membership and Non-Membership Functions of Intuitionistic Fuzzy Sets (Szmidt and Baldwin [162])

1. Due to the explanations above, from Theorem 2.1 we calculate the values of the left hand sides of the equations:

$$Pos^+(x) = \mu(x) + \pi(x) \tag{2.95}$$

and

$$Pos^-(x) = v(x) + \pi(x). \tag{2.96}$$

2. Taking into account that $\mu(x) + v(x) + \pi(x) = 1$, from (2.95)–(2.96) we obtain the values $\pi(x)$

$$Pos^+(x) + Pos^-(x) = \mu(x) + \pi(x) + v(x) + $$
$$+ \pi(x) = 1 + \pi(x) \tag{2.97}$$

$$\pi(x) = Pos^+(x) + Pos^-(x) - 1 \tag{2.98}$$

3. Knowing the values $\pi(x)$, from (2.95) and (2.96) we obtain for each x: the values $\mu(x)$, and $v(x)$.

To illustrate the above procedure we will consider a simple example showing that starting from relative frequency distributions, and using Theorem 2.1, we obtain full description of an intuitionistic fuzzy set.

Example 2.7. (Szmidt and Baldwin [162]) The task is to classify products (taking into account presence of 10 different levels of an element) as legal and illegal. Relative frequencies obtained from data for legal and illegal products are respectively

- relative frequencies $p^+(i)$ for legal products (for each i-th level of the presence of the considered element), $i = 1, \ldots, 10$

$$p^+(1) = 0., \quad p^+(2) = 0., \quad p^+(3) = 0.034,$$
$$p^+(4) = 0.165, p^+(5) = 0.301, p^+(6) = 0.301,$$
$$p^+(7) = 0.165, p^+(8) = 0.034, p^+(9) = 0.,$$
$$p^+(10) = 0. \tag{2.99}$$

- relative frequencies $p^-(i)$ for illegal products (for each i-th level of the presence of the considered element), $i = 1, \ldots, 10$

$$p^-(1) = 0.125, p^-(2) = 0.128, p^-(3) = 0.117,$$
$$p^-(4) = 0.08, \quad p^-(5) = 0.05, \quad p^-(6) = 0.05,$$
$$p^-(7) = 0.08, \quad p^-(8) = 0.117, p^-(9) = 0.128,$$
$$p^-(10) = 0.125 \tag{2.100}$$

From the data (2.99), and Theorem 2.1 we obtain possibilities $Pos^+(i)$ for legal products

$$Pos^+(1) = 0., \quad Pos^+(2) = 0., \quad Pos^+(3) = 0.205,$$
$$Pos^+(4) = 0.727, Pos^+(5) = 1., \quad Pos^+(6) = 1.,$$
$$Pos^+(7) = 0.727, Pos^+(8) = 0.205, Pos^+(9) = 0.,$$
$$Pos^+(10) = 0. \tag{2.101}$$

From the data (2.100) and Theorem 2.1 we obtain possibilities $Pos^-(i)$ for illegal products

$$Pos^-(1) = 1., \quad Pos^-(2) = 1., \quad Pos^-(3) = 0.961,$$
$$Pos^-(4) = 0.737, Pos^-(5) = 0.503, Pos^-(6) = 0.503,$$
$$Pos^-(7) = 0.737, Pos^-(8) = 0.961, Pos^-(9) = 1.,$$
$$Pos^-(10) = 1. \tag{2.102}$$

From (2.101), (2.102), and (2.98), we obtain the following values of $\pi(i)$

$$\pi(1) = 0., \quad \pi(2) = 0., \quad \pi(3) = 0.166,$$
$$\pi(4) = 0.464, \pi(5) = 0.503, \pi(6) = 0.503,$$
$$\pi(7) = 0.464, \pi(8) = 0.166, \pi(9) = 0.,$$
$$\pi(10) = 0. \tag{2.103}$$

Thus, (2.101) and (2.103) give $\mu(i)$

$$\mu(1) = 0., \quad \mu(2) = 0., \quad \mu(3) = 0.039,$$
$$\mu(4) = 0.263, \mu(5) = 0.497, \mu(6) = 0.497,$$
$$\mu(7) = 0.263, \mu(8) = 0.039, \mu(9) = 0.,$$
$$\mu(10) = 0. \tag{2.104}$$

next, from(2.102) and (2.103) we obtain $v(i)$

$$v(1) = 1., \quad v(2) = 1., \quad v(3) = 0.795,$$
$$v(4) = 0.273, v(5) = 0., \quad v(6) = 0.,$$
$$v(7) = 0.273, v(8) = 0.795, v(9) = 1.,$$
$$v(10) = 1. \tag{2.105}$$

In result, making use of relative frequencies we have obtained the values μ (2.104), v (2.105), and π (2.103) characterizing the corresponding intuitionistic fuzzy set.

Finally, we would like to emphasize the decisive difference as far as the approach discussed above is concerned, and the incorrect method of expressing an intuitionistic fuzzy set via two fuzzy sets constructed in a such way that membership values of the first fuzzy set are treated as the membership values of the intuitionistic fuzzy set, whereas membership values of the second fuzzy set are treated as the non-membership values of the same intuitionistic fuzzy set.

It is worth mentioning that the approach presented here was successfully applied for benchmark data from

UCI Machine Learning Repository (www.ics.uci.edu/ mlearn/).

The resulting intuitionistic fuzzy models were applied in:

- constructing a classifier for imbalanced and overlapping classes (cf. Szmidt and Kukier [229], [230], [231]),
- constructing intuitionistic fuzzy trees (Bujnowski [42]),

and were used for testing new approaches of calculating:

- Pearson correlation coefficient (Szmidt and Kacprzyk [211], [224]), (Szmidt et al. [220], [221], [226]),
- Spearman correlation coefficient (Szmidt and Kacprzyk [212]),
- Kendall correlation coefficient (Szmidt and Kacprzyk [222], [223]),
- Principal Component Analysis (Szmidt and Kacprzyk [225], Szmidt et al. [227]),
- ranking procedures (Szmidt and Kacprzyk [197], [198], [205], [206], [208], [214]).

2.6 Concluding Remarks

We have recalled basic definitions, and gave short introduction concerning the crisp sets, the fuzzy sets, and the intuitionistic fuzzy sets.

Two geometrical representations of the intuitionistic fuzzy sets were presented.

The interrelations among the crisp sets, the fuzzy sets, and the intuitionistic fuzzy sets were discussed.

Finally, two approaches to derivation of the intuitionistic fuzzy sets from data were presented. The first approach is by asking the experts. The second approach is automatic – starting from relative frequency distributions.

Both approaches seem to be useful. But the second one – the automatic, and mathematically justified method of deriving the intuitionistic fuzzy sets from data seems to be especially important in the context of analyzing information in big data bases. The approach has been proved to be useful in the context of widely used benchmark data.

Chapter 3
Distances

Abstract. In many theoretical and practical issues we face the following problem. Having two sets in the same universe, we want to calculate a difference between them exemplified by a distance. In this Chapter we consider distances between the intuitionistic fuzzy sets in two ways: while using the two term intuitionistic fuzzy set representation (membership values and non-membership values only are taken into account), and the three term intuitionistic fuzzy set representation (membership values, non-membership values, and hesitation margins are taken into account). We discuss norms and metrics for both types of representations. Both types are correct from the mathematical point of view but, in the practical perspective, the three term approach seems to be more justified. We discuss the problem in detail, considering its analytical, and geometrical aspects. We also show some problems with the Hausdorff distance, while the Hamming metric is applied when using the two term intuitionistic fuzzy set representation. We also show that the method of calculating the Hausdorff distances, which is correct for the interval-valued fuzzy sets, does not work for the intuitionistic fuzzy sets. Finally, we show the usefulness of the three term distances in a measure for ranking the intuitionistic fuzzy alternatives.

3.1 Basic Definitions

Definition 3.1. A distance on a set X is a positive function d (also called metric) from pairs of elements of X to the set R^+ of non-negative real numbers with the following properties, valid for all $x_1, x_2, x_3 \in X$:

1. $d(x_1, x_1) = 0$ (reflexivity);
2. $d(x_1, x_2) = 0$ if and only if $x_1 = x_2$ (separability);
3. $d(x_1, x_2) = d(x_2, x_1)$ (symmetry);
4. $d(x_1, x_3) \leq d(x_1, x_2) + d(x_2, x_3)$ (triangle inequality).

The pair (X, d) is called metric space.

E. Szmidt, *Distances and Similarities in Intuitionistic Fuzzy Sets,*
Studies in Fuzziness and Soft Computing 307,
DOI: 10.1007/978-3-319-01640-5_3, © Springer International Publishing Switzerland 2014

If a measure fulfills requirements 1, 3 and 4, it is called a pseudometric (separability does not hold).

A semimetric is defined with requirements 1, 2 and 3 (triangle inequality does not need to be satisfied).

A semi-pseudometric satisfies 1 and 3 only.

If a set of elements is identified with a vector space, the most known distances correspond to norms.

A norm of a vector corresponds in a sense to the absolute value (magnitude) of numbers.

Definition 3.2. (Bronshtein [41])

We assign a real positive number $\| x \|$ (*Norm* $\mathbf{x}$) to the vector $\mathbf{x}$. A number $\| x \|$, in order to be a norm, must satisfy the norm axioms which for any vector $\mathbf{x} \in \mathbf{R^n}$ are the following:

1. $\| \mathbf{x} \| \geq 0$ for every $\mathbf{x}$;

2. $\| \mathbf{x} \| = 0$ if and only if $\mathbf{x} = 0$;

3. $\| \alpha\mathbf{x} \| = |\alpha| \, \| \mathbf{x} \|$ for every $\mathbf{x}$ and every real number α;

4. $\| \mathbf{x} + \mathbf{y} \| \leq \| \mathbf{x} \| + \| \mathbf{y} \|$ for every $\mathbf{x}$ and $\mathbf{y}$.

Concrete norms are defined in many different ways.

If $\mathbf{x} = (x_1, x_2, \ldots, x_n)^T$ is a real vector of n dimensions, i.e., $\mathbf{x} \in \mathbf{R}^n$ then the most often used vector norms are:

Euclidean norm

$$\| \mathbf{x} \| = \| \mathbf{x} \|_2 = \sqrt{\sum_i^n x_i^2}. \tag{3.1}$$

Supremum or **Uniform Norm**

$$\| \mathbf{x} \| = \| \mathbf{x} \|_\infty = \max_{1 \leq i \leq n} |x_i|. \tag{3.2}$$

Sum Norm

$$\| \mathbf{x} \| = \| \mathbf{x} \|_1 = \sum_i^n |x_i|. \tag{3.3}$$

In applications, the so called l_r-norms and l^r-norms are often used, defined as follows.

Definition 3.3. For a vector $\mathbf{x} = (x_1, \ldots, x_n) \in \mathbf{R}^n$, its l_r-norm, where r is a real number ≥ 1, is:

$$l_r(\mathbf{x}) = \parallel \mathbf{x} \parallel_r = \left(\sum_i^n |x_i|^r \right)^{\frac{1}{r}}. \tag{3.4}$$

Slight modification of the axioms in Definition 3.2 makes it possible to define i-th power of l_r-norm, i.e., the l^r-norm:

$$l^r(\mathbf{x}) = \parallel \mathbf{x} \parallel^r = \sum_i^n |x_i|^r. \tag{3.5}$$

Euclidean norm (3.1) is a special case of the l_r-norm (3.4).

Sum norm (3.3) is a special case of the l^r-norm (3.5).

Norms on vector spaces correspond to certain metrics, i.e., every norm determines a metric, and some metrics determine a norm.

Given a normed vector space $(X \parallel \cdot \parallel)$ we can define a metric on X by $d(x,y) = \parallel x - y \parallel$. The metric d is said to be induced by the norm $\parallel \cdot \parallel$.

We give below the most often used metrics $d_{i,j} = d(y_i, y_j)$ of vectors y_i and y_j having one extreme at the origin of the coordinate axes.

- Manhattan distance

$$d_{i,j} = \sum_{k=1}^{n} |y_{ik} - y_{jk}| \tag{3.6}$$

- Euclidean distance

$$d_{i,j} = \sqrt{\sum_{k=1}^{n} (y_{ik} - y_{jk})^2} \tag{3.7}$$

- Minkowski distance

$$d_{i,j} = \left(\sum_{k=1}^{n} |y_{ik} - y_{jk}|^p \right)^{\frac{1}{p}} \tag{3.8}$$

Minkowski distance is induced by the norm l_r, namely, for $r = 1$ (3.8) it becomes Manhattan distance (city block distance); for $r = 2$ it is equivalent to the Euclidean distance.
- Chebyshev distance

$$d_{i,j} = \max_k |y_{ik} - y_{jk}| \tag{3.9}$$

Chebyshev distance is also induced by the norm l_r when $r \to \infty$.
- Canberra distance

$$d_{i,j} = \sum_{k=1}^{n} \frac{|y_{ik} - y_{jk}|}{|y_{ik}| + |y_{jk}|} \tag{3.10}$$

Canberra distance (Lance and Williams [103]) is similar to the Manhattan distance. Each component of the sum (3.10) belongs to the interval $[0,1]$. If y_{ik} or y_{jk} is equal to 0, the respective component of the sum (3.10) is equal to 1 regardless of the value of the other component. The distance is rather sensitive to small changes when both components tend to zero (Apolloni et al. [2]). For practical purposes we assume value of 0 for both coordinates equal 0 (Emran and Ye [69]).

- Sorensen distance (also known as Bray Curtis)

$$d_{i,j} = \frac{\sum_{k=1}^{n} |y_{ik} - y_{jk}|}{\sum_{k=1}^{n} (y_{ik} + y_{jk})} \tag{3.11}$$

Sorensen distance (Bray and Curtis [40]) is a modified Manhattan distance. Sorensen distance value is between zero and one if all coordinates are positive. If denominator in (3.11) is zero, Sorensen distance is undefined.
- Mahalanobis distance

$$d_{i,j} = \sqrt{\sum_{k=1}^{n} (y_{ik} - y_{jk}) S^{-1} (y_{ik} - y_{jk})} \tag{3.12}$$

where S is a covariance matrix. Mahalanobis distance ([121]) can also be defined as a dissimilarity measure between two random vectors of the same distribution with the covariance matrix S. If S is the identity matrix, the Mahalanobis distance (3.12) ([121]) is equal to the Euclidean distance (3.7). For a diagonal covariance matrix S, the Mahalanobis distance (3.12) reduces to the normalized Euclidean distance. Mahalanobis distance is used in classification methods and cluster analysis (McLachlan [122]).

The above distances d_{ij} play often a role of measures corresponding to similarity measures s_{ij}, i.e., $s_{ij} = 1 - d_{ij}$, and are widely used for solving real problems (cf. e.g., Bray and Curtis [40], Apolloni et al. [2], McLachlan [122], Emran and Ye [69], Lance and Williams [103], Krebs [102], Hublek [84], Wolda [246], Clarke et al. [54], Field et al. [71]).

In vector spaces also other similarity measures are used, for example:

- Angular Separation

$$s_{i,j} = \frac{\sum_{k=1}^{n} y_{ik} y_{jk}}{\sqrt{\sum_{k=1}^{n} (y_{ik})^2 \sum_{k=1}^{n} (y_{jk})^2}} \tag{3.13}$$

Angular separation represents cosine between two vectors. The values of (3.13) belong to the interval [-1, 1]. The higher the values of (3.13), the more similar the vectors considered. If denominator is equal to zero, we assume 0 for angular separation.
- Correlation Coefficient

$$s_{i,j} = \frac{\sum_{k=1}^{n} (y_{ik} - \bar{y}_i)(y_{jk} - \bar{y}_j)}{\sqrt{\sum_{k=1}^{n} (y_{ik} - \bar{y}_i)^2 \sum_{k=1}^{n} (y_{jk} - \bar{y}_j)^2}} \tag{3.14}$$

Correlation coefficient is a standarized angular separation resulting from centering the coordinates with respect to the mean values.

3.2   Norms and Metrics Over the Intuitionistic Fuzzy Sets or their Elements – The Two Term Approach

It is worth stressing that this section will not be devoted to the usual set-theoretic properties of the intuitionistic fuzzy sets (i.e. the properties which are a direct result of the fact that the intuitionistic fuzzy sets are sets in the sense of set theory). For example, in a metric space X, one can study the metric properties of the intuitionistic fuzzy sets over X. This can be done directly by topological methods (cf. e.g., Schwartz [149]) without paying attention to the essential properties of the intuitionistic fuzzy sets. On the other hand, all intuitionistic fuzzy sets (and hence, all fuzzy sets) over a fixed universe X generate a space (in the sense of Schwartz [149]), but with a special metric (cf., e.g., Kaufmann [99]) which is not related to the elements of X but to the values of the functions μ_A and ν_A, defined for these elements.

We should have in mind that a "norm" of a given intuitionistic fuzzy element is actually not a norm in the sense of Schwartz [149], but rather a "pseudo-norm", assigning a number to every element $x \in X$. This number depends on the values of the functions μ_A and ν_A (which are calculated for this element).

In other words, the essential conditions for a norm, i.e.:

$$\|x\| = 0 \quad \text{iff} \quad x = 0, \tag{3.15}$$

and

$$\|x\| = \|y\| \quad \text{iff} \quad x = y, \tag{3.16}$$

are not fulfilled here.

Instead of (3.15)–(3.16), the following conditions hold:

$$\|x\| = \|y\| \text{ iff } \mu_A(x) = \mu_A(y) \tag{3.17}$$

and

$$\nu_A(x) = \nu_A(y). \tag{3.18}$$

For any element $x \in X$ in every fuzzy set over X, the value of $\mu_A(x)$ plays the role of a norm (more precisely, a pseudo-norm).

In the case of the intuitionistic fuzzy sets, the presence of the second functional component, namely, the function ν_A gives rise to different possibilities for the definition of a norm (in the sense of a pseudo-norm) over the subsets and the elements of a given universe X (Atanassov [8], [11], [15], [22]).

Definition 3.4. The first norm given by Atanassov (Atanassov [15]) for every $x \in X$ with respect to a fixed set $A \subset X$ is

$$\sigma_{1,A}(x) = \mu_A(x) + \nu_A(x). \tag{3.19}$$

Norm (3.19) represents the degree of *definiteness* (Atanassov [15]) of the element x. From

$$\pi_A(x) = 1 - \mu_A(x) - \nu_A(x)$$

we can express (3.19) as

$$\sigma_{1,A}(x) = 1 - \pi_A(x).$$

For every two intuitionistic fuzzy sets A and B, and for every $x \in X$ the following properties of (3.19) hold (Atanassov [15]):

$$\sigma_{1,\overline{A}}(x) = \sigma_{1,A}(x), \tag{3.20}$$

$$\sigma_{1,A \cap B}(x) \geq \min(\sigma_{1,A}(x), \sigma_{1,B}(x)), \tag{3.21}$$

$$\sigma_{1,A \cup B}(x) \leq \max(\sigma_{1,A}(x), \sigma_{1,B}(x)), \tag{3.22}$$

$$\sigma_{1,A+B}(x) \leq 1, \tag{3.23}$$

$$\sigma_{1,A.B}(x) \leq 1, \tag{3.24}$$

$$\sigma_{1,A@B}(x) = \frac{(\sigma_{1,A}(x) + \sigma_{1,B}(x))}{2}, \tag{3.25}$$

$$\sigma_{1,A\$B}(x) \leq \frac{(\sigma_{1,A}(x) + \sigma_{1,B}(x))}{2}, \tag{3.26}$$

$$\sigma_{1,A \bowtie B}(x) \leq \max(\sigma_{1,A}(x), \sigma_{1,B}(x)), \tag{3.27}$$

$$\sigma_{1,A*B}(x) \geq \frac{\max(\sigma_{1,A}(x), \sigma_{1,B}(x))}{2}, \tag{3.28}$$

$$\sigma_{1,\Box A}(x) = 1, \tag{3.29}$$

$$\sigma_{1,\Diamond A}(x) = 1, \tag{3.30}$$

$$\sigma_{1,C(A)}(x) \geq \max_{x \in X} \sigma_{1,A}(x), \tag{3.31}$$

$$\sigma_{1,I(A)}(x) \leq \min_{x \in X} \sigma_{1,A}(x), \tag{3.32}$$

$$\sigma_{1,D_\alpha}(x) = 1 \tag{3.33}$$

for every $\alpha \in [0,1]$,

$$\sigma_{1,F_{\alpha,\beta}}(x) = \alpha + \beta + (1 - \alpha - \beta).\sigma_{1,A}(x) \tag{3.34}$$

for every $\alpha, \beta \in [0,1]$ and $\alpha + \beta \le 1$,

$$\sigma_{1,G_{\alpha,\beta}}(x) \le \sigma_{1,A}(x), \tag{3.35}$$

for every $\alpha, \beta \in [0,1]$,

$$\sigma_{1,H_{\alpha,\beta}}(x) \le \beta + (\alpha + \beta).\sigma_{1,A}(x), \tag{3.36}$$

for every $\alpha, \beta \in [0,1]$,

$$\sigma_{1,H^*_{\alpha,\beta}(A)}(x) \le \beta + (1 - \beta).\sigma_{1,A}(x), \tag{3.37}$$

for every $\alpha, \beta \in [0,1]$,

$$\sigma_{1,J_{\alpha,\beta}}(x) \le \alpha + (\alpha + \beta).\sigma_{1,A}(x), \tag{3.38}$$

for every $\alpha, \beta \in [0,1]$,

$$\sigma_{1,J^*_{\alpha,\beta}(A)}(x) \le \alpha + (1 - \alpha).\sigma_{1,A}(x), \tag{3.39}$$

for every $\alpha, \beta \in [0,1]$,

$$\sigma_{1,!A}(x) \ge 0, \tag{3.40}$$
$$\sigma_{1,?A}(x) \ge 0, \tag{3.41}$$

$$\sigma_{1,K_\alpha}(x) \ge 0, \tag{3.42}$$

for every $\alpha \in [0,1]$,

$$\sigma_{1,L_\alpha}(x) \ge 0, \tag{3.43}$$

for every $\alpha \in [0,1]$,

$$\sigma_{1,P_{\alpha,\beta}}(x) \ge 0, \tag{3.44}$$

for every $\alpha, \beta \in [0,1]$ and $\alpha + \beta \le 1$,

$$\sigma_{1,Q_{\alpha,\beta}}(x) \ge 0, \tag{3.45}$$

for every $\alpha, \beta \in [0,1]$ and $\alpha + \beta \le 1$.

Definition 3.5. Another norm for every $x \in X$, with respect to a fixed $A \subset X$, is defined as follows (Atanassov [15]):

$$\sigma_{2,A}(x) = \sqrt{(\mu_A(x)^2 + v_A(x)^2)}. \tag{3.46}$$

The norms σ_1 (3.19) and σ_2 (3.46) are analogous to the basic classical types of norms.

For the norm σ_2 (3.46), the following properties are fulfilled for every two intuitionistic fuzzy sets A and B, and for every $x \in X$ (Atanassov [15]):

$$\sigma_{2,\bar{A}}(x) = \sigma_{2,A}(x), \tag{3.47}$$

$$\sigma_{2,A \cap B}(x) \geq \min(\sigma_{2,A}(x), \sigma_{2,B}(x)), \tag{3.48}$$

$$\sigma_{2,A \cup B}(x) \leq \max(\sigma_{2,A}(x), \sigma_{2,B}(x)), \tag{3.49}$$

$$\sigma_{2,A+B}(x) \leq 1, \tag{3.50}$$

$$\sigma_{2,A.B}(x) \leq 1, \tag{3.51}$$

$$\sigma_{2,A@B}(x) \leq \frac{1}{\sqrt{2}} \cdot (\sigma_{2,A}(x) + \sigma_{2,B}(x)), \tag{3.52}$$

$$\sigma_{2,A\$B}(x) \leq \sqrt{\sigma_{2,A}(x) \cdot \sigma_{2,B}(x)}, \tag{3.53}$$

$$\sigma_{2,A \bowtie B}(x) \geq \min(\sigma_{2,A}(x), \sigma_{2,B}(x)), \tag{3.54}$$

$$\sigma_{2,A*B}(x) \geq \max(\sigma_{2,A}(x), \sigma_{2,B}(x))/2, \tag{3.55}$$

$$\sigma_{2,\square A}(x) \leq 1, \tag{3.56}$$

$$\sigma_{2,\Diamond A}(x) \leq 1, \tag{3.57}$$

$$\sigma_{2,CA}(x) \leq \max_{x \in X} \sigma_{2,A}(x), \tag{3.58}$$

$$\sigma_{2,IA}(x) \geq \min_{x \in X} \sigma_{2,A}(x), \tag{3.59}$$

$$\sigma_{2,D_\alpha}(x) \geq \sigma_{2,A}(x), \tag{3.60}$$

for every $\alpha \in [0,1]$,

$$\sigma_{2,F_{\alpha,\beta}}(x) \geq \sigma_{2,A}(x), \tag{3.61}$$

for every $\alpha, \beta \in [0,1]$ such that $\alpha + \beta \leq 1$,

$$\sigma_{2,G_{\alpha,\beta}}(x) \leq \sigma_{2,A}(x),\qquad(3.62)$$

for every $\alpha, \beta \in [0,1]$,

$$\sigma_{2,H_{\alpha,\beta}}(x) \geq \alpha.\sigma_{2,A}(x),\qquad(3.63)$$

for every $\alpha, \beta \in [0,1]$,

$$\sigma_{2,H^*_{\alpha,\beta}(A)}(x) \geq \alpha.\sigma_{2,A}(x),\qquad(3.64)$$

for every $\alpha, \beta \in [0,1]$,

$$\sigma_{2,J_{\alpha,\beta}}(x) \geq \beta.\sigma_{2,A}(x),\qquad(3.65)$$

for every $\alpha, \beta \in [0,1]$,

$$\sigma_{2,J^*_{\alpha,\beta}(A)}(x) \geq \beta.\sigma_{2,A}(x),\qquad(3.66)$$

for every $\alpha, \beta \in [0,1]$,

$$\sigma_{2,!A}(x) \geq \frac{1}{2},\qquad(3.67)$$
$$\sigma_{2,?A}(x) \geq \frac{1}{2},\qquad(3.68)$$
$$\sigma_{2,K_\alpha(A)}(x) \geq \alpha,\qquad(3.69)$$

for every $\alpha \in [0,1]$,

$$\sigma_{2,L_\alpha(A)}(x) \geq \alpha,\qquad(3.70)$$

for every $\alpha \in [0,1]$,

$$\sigma_{2,P_{\alpha,\beta}(A)}(x) \geq \alpha,\qquad(3.71)$$

for every $\alpha, \beta \in [0,1]$ and $\alpha + \beta \leq 1$,

$$\sigma_{2,Q_{\alpha,\beta}(A)}(x) \geq \beta,\qquad(3.72)$$

for every $\alpha, \beta \in [0,1]$ and $\alpha + \beta \leq 1$.

Definition 3.6. Tanev [235]) defined the third norm over the elements of a given intuitionistic fuzzy set A as:

$$\sigma_{3,A}(x) = \frac{\mu_A(x) + 1 - \nu_A(x)}{2}.\qquad(3.73)$$

The properties of (3.73) are similar to the properties of the first norm (3.19), and the second one, (3.46).

Some other discrete norms introduced by Atanassov [15] are presented in Definition 3.7.

Definition 3.7. For a given finite universe X and for a given intuitionistic fuzzy set A, we have the following discrete norms (Atanassov [15]):

$$n_\mu(A) = \sum_{x \in X} \mu_A(x), \qquad (3.74)$$

$$n_\nu(A) = \sum_{x \in X} \nu_A(x), \qquad (3.75)$$

$$n_\pi(A) = \sum_{x \in X} \pi_A(x). \qquad (3.76)$$

The above norms (3.74)–(3.76) can be extended to continuous norms by replacing the sum in (3.74)–(3.76) by an integral over X.

After normalizing the norms (3.74)–(3.76) on the interval $[0,1]$, we obtain for a given finite universe X and for a given intuitionistic fuzzy set A, the following normalized discrete norms (Atanassov [15]):

- corresponding to the norm "$n_\mu(A)$" (3.74)

$$n_\mu^*(A) = \frac{1}{card(X)} \sum_{x \in X} \mu_A(x), \qquad (3.77)$$

- corresponding to the norm "$n_\nu(A)$" (3.76)

$$n_\nu^*(A) = \frac{1}{card(X)} \sum_{x \in X} \nu_A(x), \qquad (3.78)$$

- corresponding to the norm "$n_\pi(A)$" (3.76)

$$n_\pi^*(A) = \frac{1}{card(X)} \sum_{x \in X} \pi_A(x), \qquad (3.79)$$

where $card(X)$ is the cardinality of the set X.

The above norms have similar properties.

In the theory of fuzzy sets (see e.g. Kaufmann [99]) two different types of distances are defined, generated from the following metric

$$m_A(x,y) = |\mu_A(x) - \mu_A(y)|$$

and the Hamming and Euclidean metrics coincide (Atanassov [15]).

In the case of the intuitionistic fuzzy sets these metrics are different (Atanassov [15]):

Definition 3.8. For an intuitionistic fuzzy set A the Hamming metric is defined as (Atanassov [15]):

$$h_A(x,y) = \frac{1}{2}(|\,\mu_A(x) - \mu_A(y) + \nu_A(x) - \nu_A(y)\,|). \qquad (3.80)$$

Definition 3.9. For an intuitionistic fuzzy set A the the Euclidean metric is defined as (Atanassov [15]):

$$e_A(x,y) = \sqrt{\frac{1}{2}((\mu_A(x) - \mu_A(y))^2 + (\nu_A(x) - \nu_A(y))^2)}. \qquad (3.81)$$

Under the assumption that

$$\nu_A(x) = 1 - \mu_A(x)$$

both metrics, (3.80) and (3.81), reduce to the metric m_A (x, y) (Atanassov [15]). To show that h_A and e_A are pseudo-metrics over X (in the sense of [100, 149]), it is necessary to prove that for every three elements $x, y, z \in X$ we have (Atanassov [15]):

$$h_A(x,y) + h_A(y,z) \geq h_A(x,z), \qquad (3.82)$$

$$h_A(x,y) = h_A(y,x), \qquad (3.83)$$

$$e_A(x,y) + e_A(y,z) \geq e_A(x,z), \qquad (3.84)$$

$$e_A(x,y) = e_A(y,x). \qquad (3.85)$$

As conditions (3.82) and (3.84) do not hold (Atanassov [15]) for the metrics, h_A and e_A are pseudo-metrics. The proofs of the above equalities and inequalities are trivial.

The well known types of distances for the fuzzy sets A and B are:

- the Hamming distance

$$d(A,B) = \sum_{x \in X} |\,\mu_A(x) - \mu_B(x)\,|, \qquad (3.86)$$

- the Euclidean distance

$$e(A,B) = \sqrt{\sum_{x \in X} (\mu_A(x) - \mu_B(x))^2}. \qquad (3.87)$$

The distances (3.86) and (3.87) transformed into the intuitionistic fuzzy sets, have the following respective forms (Atanassov [15]):

Definition 3.10. For two intuitionistic fuzzy sets A and B over a universe X, the Hamming distance between A and B is defined as (Atanassov [15])

$$d_{IFS(2)}(A,B) = \frac{1}{2}\sum_{x\in X} \mid \mu_A(x) - \mu_B(x)\mid + \mid \nu_A(x) - \nu_B(x)\mid, \qquad (3.88)$$

and the corresponding normalized Hamming distance is

$$l_{IFS(2)}(A,B) = \frac{1}{2n}\sum_{x\in X} \mid \mu_A(x) - \mu_B(x)\mid + \mid \nu_A(x) - \nu_B(x)\mid. \qquad (3.89)$$

Definition 3.11. For two intuitionistic fuzzy sets A and B over a universe X, the Euclidean distance between A and B is defined as (Atanassov [15])

$$e_{IFS(2)}(A,B) = \sqrt{\frac{1}{2}(\sum_{x\in X}(\mu_A(x) - \mu_B(x))^2 + (\nu_A(x) - \nu_B(x))^2)}, \qquad (3.90)$$

and the corresponding normalized Euclidean distance is

$$q_{IFS(2)}(A,B) = \sqrt{\frac{1}{2n}(\sum_{x\in X}(\mu_A(x) - \mu_B(x))^2 + (\nu_A(x) - \nu_B(x))^2)}. \qquad (3.91)$$

Distances (3.88)–(3.91) correspond to the two term intuitionistic fuzzy set description – i.e. the membership values and the non-membership values are taken into account only. In Section 3.3 we will discuss another form of the Hamming and Euclidean distances, using the three term description of the intuitionistic fuzzy sets (besides the membership values and the non-membership values also the hesitation margin is taken into account):

$$l^1_{IFS}(A,B) = \frac{1}{2n}\sum_{x\in E}\mid \mu_A(x) - \mu_B(x)\mid + \mid \nu_A(x) - \nu_B(x)\mid + \mid \pi_A(x) - \pi_B(x)\mid \quad (3.92)$$

and

$$q^1_{IFS}(A,B) = \sqrt{\frac{1}{2n}(\sum_{x\in E}(\mu_A(x) - \mu_B(x))^2 + (\nu_A(x) - \nu_B(x))^2) + (\pi_A(x) - \pi_B(x))^2)}. \qquad (3.93)$$

Distances (3.92)–(3.93) correspond to the three term intuitionistic fuzzy set description (membership values, non-membership values and hesitation margins are taken into account) and are useful from the point of view of practical applications.

In the next chapter we will discuss in details distances (3.88)–(3.90) and (3.92)–(3.93).

In (Atanassov [15]), other distances are also given (cf. [149]), which can be defined over the intutionistic fuzzy sets:

Definition 3.12. For two intuitionistic fuzzy sets A and B over a universe X, the following distances between A and B are defined (Atanassov [15]):

$$J_1(A,B) = \max_{x \in X} \mid \mu_A(x) - \mu_B(x) \mid, \tag{3.94}$$

$$J_2(A,B) = \max_{x \in X} \mid v_A(x) - v_B(x) \mid, \tag{3.95}$$

$$J(A,B) = \frac{1}{2} \cdot (J_1(A,B) + J_2(A,B)), \tag{3.96}$$

$$J^*(A,B) = \frac{1}{2} \cdot \max_{x \in X} (\mid \mu_A(x) - \mu_B(x) \mid + \mid v_A(x) - v_B(x) \mid). \tag{3.97}$$

It is easily seen that for every two intuitionistic fuzzy sets A and B we have (Atanassov [15]):

$$J^*(A,B) \le J_1(A,B) + J_2(A,B).$$

In the distance $J_1(.,.)$ (3.94) only the membership values are taken into account, and so the distance is reduced directly to the distance for fuzzy sets. On the other hand, the distances $J_2(.,.)$ (3.95), $J(.,.)$ (3.96), and $J^*(.,.)$ (3.97) make use of both membership and non-membership values, and thus they do not reduce to the distances for fuzzy sets.

Atanassov [21], [22] introduced also norms following one of the most important ideas of Georg Cantor in set theory, calling the norms "*Cantor's intuitionistic fuzzy norms*". Cantor's intuitionistic fuzzy norms are substantially different from the Euclidean and Hamming norms, existing in fuzzy set theory.

Let $x \in X$ be fixed universe and let

$$\mu_A(x) = 0.a_1 a_2 ...$$

$$v_A(x) = 0.b_1 b_2 ...$$

Next, Atanassov [22] bijectively constructed the numbers:

$$||x||_{\mu,v} = 0, a_1 b_1 a_2 b_2 ...$$

and

$$||x||_{v,\mu} = 0, b_1 a_1 b_2 a_2 ...$$

and noticed that the following properties hold for these numbers:

1. $||x||_{\mu,v}, ||x||_{v,\mu} \in [0,1]$
2. having both numbers it is possible to reconstruct directly the numbers $\mu_A(x)$ and $v_A(x)$.

The numbers $||x||_{\mu,\nu}$ and $||x||_{\nu,\mu}$ were called by Atanassov [22] Cantor norms of element $x \in X$.

Atanassov [22] denotes these norms by $||x||_{2,\mu,\nu}$ and $||x||_{2,\nu,\mu}$ in order to stress that they correspond to the two term intuitionistic fuzzy set description.

On the other hand, for the three term intuitionistic fuzzy set description introduced by Szmidt and Kacprzyk, for point x we have (Atanassov [22]):

$$\mu_A(x) = 0.a_1a_2...$$

$$\nu_A(x) = 0.b_1b_2...$$

$$\pi_A(x) = 0.c_1c_2...$$

with the condition: $\mu_A(x) + \nu_A(x) + \pi_A(x) = 1$. (Atanassov [22]) introduced six different Cantor norms:

$$||x||_{3,\mu,\nu,\pi} = 0.a_1b_1c_1a_2b_2c_2...,$$

$$||x||_{3,\mu,\pi,\nu} = 0.a_1c_1b_1a_2c_2b_2...,$$

$$||x||_{3,\nu,\mu,\pi} = 0.b_1a_1c_1b_2a_2c_2...,$$

$$||x||_{3,\nu,\pi,\mu} = 0.b_1c_1a_1b_2c_2a_2...,$$

$$||x||_{3,\pi,\mu,\nu} = 0.c_1a_1b_1c_2a_2b_2...,$$

$$||x||_{3,\pi,\nu,\mu} = 0.c_1b_1a_1c_2b_2a_2....$$

For the above three term Cantor norms it is possible, as previously, to reconstruct bijectively the three degrees of element $x \in X$.

3.3 Distances between the Intuitionistic Fuzzy Sets – The Three Term Approach

In this section we recall some new definitions of distances between intuitionistic fuzzy sets (Szmidt and Kacprzyk [171]). By taking into account the three term characterization of the intuitionistic fuzzy sets, and following the basic line of reasoning on which the definition of distances between the fuzzy sets is based, we define four basic distances between the intuitionistic fuzzy sets: Hamming distance, normalized Hamming distance, Euclidean distance, and normalized Euclidean distance. While deriving these distances a convenient geometric interpretation of the intuitionistic fuzzy sets is employed. It is shown that the definitions proposed are consistent with their counterparts traditionally used for the fuzzy sets, and that the consistency is

ensured only under the condition that all three parameters characterizing the intuitionistic fuzzy sets are taken into account.

We will first reconsider some better known distances for the fuzzy sets in an
intuitionistic setting, and then extend those distances to the intuitionistic fuzzy sets.

3.3.1   Distances between the Fuzzy Sets

The most widely used distances for fuzzy sets A', B' in $X = \{x_1, x_2, ..., x_n\}$ are
(Kacprzyk, 1997):

- the Hamming distance $d(A', B')$

$$d(A', B') = \sum_{i=1}^{n} |\mu_{A'}(x_i) - \mu_{B'}(x_i)| \qquad (3.98)$$

- the normalized Hamming distance $l(A', B')$:

$$l(A', B') = \frac{1}{n} \sum_{i=1}^{n} |\mu_{A'}(x_i) - \mu_{B'}(x_i)| \qquad (3.99)$$

- the Euclidean distance $e(A', B')$:

$$e(A', B') = \sqrt{\sum_{i=1}^{n} \left(\mu_{A'}(x_i) - \mu_{B'}(x_i)\right)^2} \qquad (3.100)$$

- the normalized Euclidean distance $q(A', B')$:

$$q(A', B') = \sqrt{\frac{1}{n} \sum_{i=1}^{n} \left(\mu_{A'}(x_i) - \mu_{B'}(x_i)\right)^2} \qquad (3.101)$$

It is worth mentioning that in the above formulas, (3.98)-(3.101), only the membership functions are present. It is due to the fact that for a fuzzy set, $\mu(x_i) + \nu(x_i) = 1$.

In Chapter 2, we have introduced for a fuzzy set A' in X an equivalent intuitionistic-type representation given as

$$A' = \{< x, \mu_{A'}(x), 1 - \mu_{A'}(x) > / x \in X\}.$$

The above representation will be employed while rewriting the distances (3.98)-
(3.101).

So, first, taking into account an intuitionistic-type representation of a fuzzy set,
we can express the very essence of the Hamming distance by putting

$$d'(A',B') = \sum_{i=1}^{n} \left(\left| \mu_{A'}(x_i) - \mu_{B'}(x_i) \right| + \left| v_{A'}(x_i) - v_{B'}(x_i) \right| \right) =$$

$$= \sum_{i=1}^{n} \left(\left| \mu_{A'}(x_i) - \mu_{B'}(x_i) \right| + \left| 1 - \mu_{A'}(x_i) - 1 + \mu_{B'}(x_i) \right| \right) =$$

$$= 2 \sum_{i=1}^{n} \left| \mu_{A'}(x_i) - \mu_{B'}(x_i) \right| = 2d(A',B') \tag{3.102}$$

i.e. the Hamming distance in an intuitionistic-type representation of the fuzzy sets is twice the Hamming distance between fuzzy sets calculated in a standard way, (3.98).

Similarly, the normalized Hamming distance $l'(A',B')$, when we take into account an intuitionistic-type representation of a fuzzy set, is in turn equal to

$$l'(A',B') = \frac{1}{n} \cdot d'(A',B') = \frac{2}{n} \sum_{i=1}^{n} \left| \mu_{A'}(x_i) - \mu_{B'}(x_i) \right| \tag{3.103}$$

i.e. the result of (3.103) is equal the well known normalized Hamming distance (3.99) between fuzzy sets, multiplied by two.

Then, by the same line of reasoning, the Euclidean distance, taking into account an intuitionistic-type representation of a fuzzy set, is equal to

$$e'(A',B') = \sqrt{\sum_{i=1}^{n} \left(\mu_{A'}(x_i) - \mu_{B'}(x_i) \right)^2 + \left(v_{A'}(x_i) - v_{B'}(x_i) \right)^2} =$$

$$= \sqrt{\sum_{i=1}^{n} \left(\mu_{A'}(x_i) - \mu_{B'}(x_i) \right)^2 + \left(1 - \mu_{A'}(x_i) - 1 + \mu_{B'}(x_i) \right)^2} =$$

$$= \sqrt{2 \sum_{i=1}^{n} \left(\mu_{A'}(x_i) - \mu_{B'}(x_i) \right)^2} \tag{3.104}$$

i.e. it is just multiplied by $\sqrt{2}$ Euclidean distance for the usual representation of fuzzy sets given by (3.100).

The normalized Euclidean distance $q'(A',B')$ considering the intuitionistic-type representation of a fuzzy set is equal to

$$q'(A',B') = \sqrt{\frac{1}{n} \cdot e'(A',B')} = \sqrt{\frac{2}{n} \sum_{i=1}^{n} \left(\mu_{A'}(x_i) - \mu_{B'}(x_i) \right)^2} \tag{3.105}$$

so again the result of (3.105) is the expression from (3.101) multiplied by $\sqrt{2}$.

Example 3.1. (Szmidt and Kacprzyk [171]) For simplicity we consider "degenerate" fuzzy sets M,N,L,K,P in $X = \{1\}$. Complete description of each of them is given by $A = (\mu_A, v_A)/1$, namely:

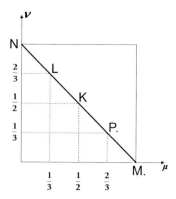

Fig. 3.1 A geometrical interpretation of one-element fuzzy sets considered in Example 3.1

$$M = (1,0)/1, \quad N = (0,1)/1, \quad L = (\frac{1}{3}, \frac{2}{3})/1, \quad P = (\frac{2}{3}, \frac{1}{3})/1, \quad K = (\frac{1}{2}, \frac{1}{2})/1$$

Figure 3.1 gives a geometrical interpretation of these one-element fuzzy sets.

First, let us calculate the Euclidean distances between the fuzzy sets using their "normal" representation (i.e., taking into account the membership values only) as in (3.100)

$$e(L,P) = \sqrt{(\frac{1}{3} - \frac{2}{3})^2} = \frac{1}{3} \tag{3.106}$$

$$e(L,K) = \sqrt{(\frac{1}{3} - \frac{1}{2})^2} = \frac{1}{6} \tag{3.107}$$

$$e(P,K) = \sqrt{(\frac{2}{3} - \frac{1}{2})^2} = \frac{1}{6} \tag{3.108}$$

$$e(L,M) = \sqrt{(\frac{1}{3} - 1)^2} = \frac{2}{3} \tag{3.109}$$

$$e(K,M) = \sqrt{(1 - \frac{1}{2})^2} = \frac{1}{2} \tag{3.110}$$

$$e(N,K) = \sqrt{(0 - \frac{1}{2})^2} = \frac{1}{2} \tag{3.111}$$

$$e(N,M) = \sqrt{1^2} = 1 \tag{3.112}$$

The same Euclidean distances are calculated now using the intuitionistic-type representation of fuzzy sets (3.104)

$$e'(L,P) = \sqrt{(\frac{1}{3} - \frac{2}{3})^2 + (\frac{2}{3} - \frac{1}{3})^2} = \frac{\sqrt{2}}{3} \qquad (3.113)$$

$$e'(L,K) = \sqrt{(\frac{1}{3} - \frac{1}{2})^2 + (\frac{2}{3} - \frac{1}{2})^2} = \frac{\sqrt{2}}{6} \qquad (3.114)$$

$$e'(P,K) = \sqrt{(\frac{2}{3} - \frac{1}{2})^2 + (\frac{1}{3} - \frac{1}{2})^2} = \frac{\sqrt{2}}{6} \qquad (3.115)$$

$$e'(L,M) = \sqrt{(\frac{1}{3} - 1)^2 + (\frac{2}{3} - 0)^2} = \frac{2\sqrt{2}}{3} \qquad (3.116)$$

$$e'(K,M) = \sqrt{(\frac{1}{2} - 1)^2 + (\frac{1}{2})^2} = \frac{\sqrt{2}}{2} \qquad (3.117)$$

$$e'(N,K) = \sqrt{(0 - \frac{1}{2})^2 + (1 - \frac{1}{2})^2} = \frac{\sqrt{2}}{2} \qquad (3.118)$$

$$e'(N,M) = \sqrt{1^2 + 1^2} = \sqrt{2} \qquad (3.119)$$

Thus, as has been already noticed, the above results are just those of (3.106)–(3.112) multiplied by the constant value equal to $\sqrt{2}$. Therefore, though the distances (3.106)–(3.112) and (3.113)–(3.119) are clearly different, their essence is the same.

Example 3.2. (Szmidt and Kacprzyk [171]) Let us consider two fuzzy sets A', B' in $X = \{1,2,3,4,5,6,7\}$. Their intuitionistic - type representation is $A' = (\mu_{A'}, v_{A'})/1$, given here as

$$A' = (0.7,\, 0.3)/1 + (0.2,\, 0.8)/2 + (0.6,\, 0.4)/4 + (0.5,\, 0.5)/5 + (1,\, 0)/6 \quad (3.120)$$

$$B' = (0.2,\, 0.8)/1 + (0.6,\, 0.4)/4 + (0.8,\, 0.2)/5 + (1,\, 0)/7 \qquad (3.121)$$

The Hamming distance $d(A', B')$, accounting only for the membership functions (3.98), is

$$d(A', B') = |0.7 - 0.2| + |0.2 - 0| + |0.6 - 0.6| + |0.5 - 0.8| + |1 - 0| + |0 - 1| = 3$$
$$(3.122)$$

while the normalized distance (3.99) $l(A', B')$ is equal to

$$l(A', B') = \frac{1}{7} \cdot d(A', B') = \frac{3}{7} = 0.43 \qquad (3.123)$$

On the other hand, when both the membership and non-membership values are taken into account [cf.(3.102)], we obtain

$$d'(A',B') = |0.7 - 0.2| + |0.3 - 0.8| + |0.2 - 0| + |0.8 - 1| + |0.6 - 0.6| +$$
$$+ |0.4 - 0.4| + |0.5 - 0.8| + |0.5 - 0.2| + |1 - 0| + |0 - 1| +$$
$$+ |0 - 1| + |1 - 0| = 6 \tag{3.124}$$

i.e. we get the value from (3.122) multiplied by two. The normalized Hamming distance (3.103) is equal to

$$l'(A',B') = \frac{1}{n}d'(A',B') = \frac{6}{7} = 0.86 \tag{3.125}$$

Let us compare the Euclidean distances obtained from (3.100) and (3.104). From (3.100) we have

$$e(A',B') = ((0.7 - 0.2)^2 + (0.2 - 0)^2 + (0.6 - 0.6)^2 + (0.5 - 0.8)^2 +$$
$$+ (1 - 0.2)^2 + (0 - 1)^2)^{\frac{1}{2}} = \sqrt{2.38} = 1.54 \tag{3.126}$$

while the counterpart normalized Euclidean distance (3.101) is

$$q(A',B') = \sqrt{\frac{1}{7} \cdot e(A,B)} = \sqrt{\frac{2.38}{7}} = 0.58 \tag{3.127}$$

From (3.104) we have the Euclidean distance, taking into account the intuitionistic-type representation of fuzzy sets, equal to

$$e'(A',B') = ((0.7 - 0.2)^2 + (0.3 - 0.8)^2 + (0.2 - 0)^2 + (0.8 - 1)^2 + (0.6 - 0.6)^2 +$$
$$+ (0.4 - 0.4)^2 + (0.5 - 0.8)^2 + (0.5 - 0.2)^2 + (1 - 0)^2 +$$
$$+ (0 - 0)^2 + (0 - 1)^2 + (1 - 0)^2)^{\frac{1}{2}} = \sqrt{4.76} = 2.18 \tag{3.128}$$

whereas the counterpart, the normalized Euclidean distance (3.105), accounting for the intuitionistic-type representation of fuzzy sets is equal to

$$q'(A',B') = \sqrt{\frac{4.76}{7}} = 0.83 \tag{3.129}$$

Suppose we modify a little bit the fuzzy set B' (making it closer to A'), i.e., these two fuzzy sets are now

$$A' = (0.7, \, 0.3)/1 + (0.2, \, 0.8)/2 + (0.6, \, 0.4)/4 + (0.5, \, 0.5)/5 + (1, \, 0)/6 \tag{3.130}$$

$$B' = (0.2, \, 0.8)/1 + (0.6, \, 0.4)/4 + (0.8, \, 0.2)/5 + (0.4, \, 0.6)/6 + (1, \, 0)/7 \tag{3.131}$$

The Hamming distance calculated with (3.98) is

$$d(A',B') = |0.7 - 0.2| + |0.2 - 0| + |0.6 - 0.6| + |0.5 - 0.8| + |1 - 0.4| + |0 - 1| =$$
$$= 2.6 \tag{3.132}$$

whereas the normalized Hamming distance (3.99) is

$$l(A',B') = \frac{1}{7} \cdot d(A',B') = \frac{2.6}{7} = 0.37 \tag{3.133}$$

From (3.102), taking into account the intuitionistic-type representation of fuzzy sets, we obtain the Hamming distance equal to

$$d'(A',B') = |0.7 - 0.2| + |0.3 - 0.8| + |0.2 - 0| + |0.8 - 1| + |0.6 - 0.6| +$$
$$+ |0.4 - 0.4| + |0.5 - 0.8| + |0.5 - 0.2| + |1 - 0.4| + |0 - 0.6| +$$
$$+ |0 - 1| + |1 - 0| = 5.2 \tag{3.134}$$

while the normalized Hamming distance (3.103) taking into account the intuitionistic-type representation of fuzzy sets, is equal to

$$l'(A',B') = \frac{1}{7} \cdot 5.2 = 0.74 \tag{3.135}$$

Let us calculate the Euclidean distances now. From (3.100) we obtain

$$e(A',B') = ((0.7 - 0.2)^2 + (0.2 - 0)^2 + (0.6 - 0.6)^2 + (0.5 - 0.8)^2 +$$
$$+ (1 - 0.4)^2 + (0 - 1)^2)^{\frac{1}{2}} = \sqrt{1.74} = 1.32 \tag{3.136}$$

while from (3.101) we get the normalized Euclidean distance

$$q(A',B') = \sqrt{\frac{1.74}{7}} = 0.5 \tag{3.137}$$

Taking into account the intuitionistic-type representation of fuzzy sets, from (3.104) we obtain the Euclidean distance

$$e'(A',B') = ((0.7 - 0.2)^2 + (0.3 - 0.8)^2 + (0.2 - 0)^2 + (0.8 - 1)^2 +$$
$$+ (0.6 - 0.6)^2 + (0.4 - 0.4)^2 + (0.5 - 0.8)^2 + (0.5 - 0.2)^2 +$$
$$+ (1 - 0.4)^2 + (0 - 0.6)^2 + (0 - 1)^2 + (1 - 0)^2)^{\frac{1}{2}} =$$
$$= \sqrt{3.48} = 1.87 \tag{3.138}$$

while the normalized Euclidean distance (3.105), taking into account the intuitionistic-type representation of fuzzy sets, is equal to

$$q'(A',B') = \sqrt{\frac{1}{7} \cdot 3.48} = 0.705 \tag{3.139}$$

As we analyze the results of Examples 3.1 and 3.2 we may notice that:

- for any fuzzy sets A' and B', when we calculate the distances between them in a standard way (3.98)–(3.101), i.e., when we take into account the membership values only, we have

$$0 \le d(A',B') \le n \tag{3.140}$$

$$0 \le l(A',B') \le 1 \tag{3.141}$$

$$0 \le e(A',B') \le \sqrt{n} \tag{3.142}$$

$$0 \le q(A',B') \le 1 \tag{3.143}$$

- for any fuzzy sets A' and B', when we calculate distances between them taking into account the intuitionistic-type representation of fuzzy sets (3.102)-(3.105), we have

$$0 \le d'(A',B') \le 2n \tag{3.144}$$

$$0 \le l'(A',B') \le 2 \tag{3.145}$$

$$0 \le e'(A',B') \le \sqrt{2n} \tag{3.146}$$

$$0 \le q'(A',B') \le \sqrt{2} \tag{3.147}$$

We would like to emphasize that it is not our purpose to introduce a new way of calculating distances for fuzzy sets. To the contrary, we have shown that the intuitionistic-type representation of fuzzy sets results in multiplying the distances by constant values only. But similar reasoning for the case of the intuitionistic fuzzy sets (i.e. omitting one of the three terms) would lead to incorrect results, as this is discussed in detail in the next section.

3.3.2 Distances between the Intuitionistic Fuzzy Sets

Following the line of reasoning presented in Section 3.3.1, we will now extend the concepts of distances to the case of the intuitionistic fuzzy sets.

The Hamming distance between two intuitionistic fuzzy sets A and B in $X = \{x_1, x_2, ..., x_n\}$ is equal to (Szmidt and Kacprzyk [171])

$$d^1_{IFS}(A,B) = \sum_{i=1}^{n} (|\mu_A(x_i) - \mu_B(x_i)| + |\nu_A(x_i) - \nu_B(x_i)| + |\pi_A(x_i) - \pi_B(x_i)|) \tag{3.148}$$

Having in mind that

$$\pi_A(x_i) = 1 - \mu_A(x_i) - \nu_A(x_i) \tag{3.149}$$

and

$$\pi_B(x_i) = 1 - \mu_B(x_i) - \nu_B(x_i) \tag{3.150}$$

we have

$$|\pi_A(x_i) - \pi_B(x_i)| = |1 - \mu_A(x_i) - \nu_A(x_i) - 1 + \mu_B(x_i) + \nu_B(x_i)| \le$$
$$\le |\mu_B(x_i) - \mu_A(x_i)| + |\nu_B(x_i) - \nu_A(x_i)| \tag{3.151}$$

From inequality (3.151) it follows that the third term in (3.148) cannot be omitted as it was in the case of fuzzy sets, for which taking into account the second term would only result in the multiplication by a constant value.

For the Euclidean distance a similar situation occurs. Namely, for intuitionistic fuzzy sets A and B in $X = \{x_1, x_2, ..., x_n\}$, by following the line of reasoning as in Section 3.3.1, their Euclidean distance is equal to (Szmidt and Kacprzyk [171])

$$e^1_{IFS}(A, B) = (\sum_{i=1}^{n} (\mu_A(x_i) - \mu_B(x_i))^2 + (\nu_A(x_i) - \nu_B(x_i))^2 +$$
$$+ (\pi_A(x_i) - \pi_B(x_i))^2)^{\frac{1}{2}} \tag{3.152}$$

Let us verify the effect of omitting the third term (π) in (3.152). Having in mind (3.149)–(3.150), we have (Szmidt and Kacprzyk [171]):

$$(\pi_A(x_i) - \pi_B(x_i))^2 = (1 - \mu_A(x_i) - \nu_A(x_i) - 1 + \mu_B(x_i) + \nu_B(x_i))^2 =$$
$$= (\mu_A(x_i) - \mu_B(x_i))^2 + (\nu_A(x_i) - \nu_B(x_i))^2 +$$
$$+ 2(\mu_A(x_i) - \mu_B(x_i))(\nu_A(x_i) - \nu_B(x_i)) \tag{3.153}$$

which means that taking into account the third term π when calculating the Euclidean distance for the intuitionistic fuzzy sets does have an influence on the final result. This is obvious, because a two-dimensional geometrical interpretation (Figure 2.2) is an orthogonal projection of a real situation presented in Figure 2.3.

Taking into account (3.149)-(3.153), in order to be more in agreement with the mathematical notion of normalization, the following distances for two intuitionistic fuzzy sets A and B in $X = \{x_1, x_2, ..., x_n\}$ are proposed (Szmidt and Kacprzyk [171])

- the Hamming distance:

$$d^1_{IFS}(A, B) = \frac{1}{2} \sum_{i=1}^{n} (|\mu_A(x_i) - \mu_B(x_i)| + |\nu_A(x_i) - \nu_B(x_i)| +$$
$$+ |\pi_A(x_i) - \pi_B(x_i)|) \tag{3.154}$$

- the Euclidean distance :

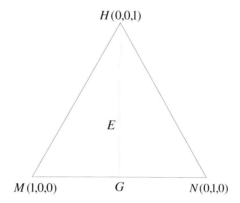

Fig. 3.2 Geometrical representation of the one-element intuitionistic fuzzy sets from Example 3.3

$$e^1_{IFS}(A,B) = (\frac{1}{2} \sum_{i=1}^{n} (\mu_A(x_i) - \mu_B(x_i))^2 + (v_A(x_i) - v_B(x_i))^2 +$$
$$+ (\pi_A(x_i) - \pi_B(x_i))^2)^{\frac{1}{2}} \qquad (3.155)$$

• the normalized Hamming distance:

$$l^1_{IFS}(A,B) = \frac{1}{2n} \sum_{i=1}^{n} (|\mu_A(x_i) - \mu_B(x_i)| + |v_A(x_i) - v_B(x_i)| +$$
$$+ |\pi_A(x_i) - \pi_B(x_i)|) \qquad (3.156)$$

• the normalized Euclidean distance:

$$q^1_{IFS}(A,B) = (\frac{1}{2n} \sum_{i=1}^{n} (\mu_A(x_i) - \mu_B(x_i))^2 + (v_A(x_i) - v_B(x_i))^2 +$$
$$+ (\pi_A(x_i) - \pi_B(x_i))^2)^{\frac{1}{2}} \qquad (3.157)$$

The above distances satisfy the conditions of the metric (cf. Kaufmann [99]).

Example 3.3. (Szmidt and Kacprzyk [171]) Let us consider for simplicity the "degenerate" intuitionistic fuzzy sets M, N, H, G, E in $X = \{1\}$. The full description of each intuitionistic fuzzy set, i.e. $A = (\mu_A, v_A, \pi_A)/1$, may be exemplified by

$$M = (1,0,0)/1, \quad N = (0,1,0)/1, \quad H = (0,0,1)/1,$$
$$G = (\frac{1}{2}, \frac{1}{2}, 0)/1, \quad E = (\frac{1}{4}, \frac{1}{4}, \frac{1}{2})/1 \qquad (3.158)$$

The geometrical interpretation of the above sets is presented in Figure 3.2.

Let us calculate the Euclidean distances between the above intuitionistic fuzzy sets omitting the third term, i.e., using the following formula:

$$e_{IFS(2)}(A,B) = \sqrt{\frac{1}{2}\sum_{i=1}^{n}(\mu_A(x_i) - \mu_B(x_i))^2 + (\nu_A(x_i) - \nu_B(x_i))^2} \qquad (3.159)$$

We obtain:

$$e_{IFS(2)}(M,H) = \sqrt{\frac{1}{2}((1-0)^2 + 0^2)} = \frac{1}{2} \qquad (3.160)$$

$$e_{IFS(2)}(N,H) = \sqrt{\frac{1}{2}(0^2 + (0-1)^2)} = \frac{1}{2} \qquad (3.161)$$

$$e_{IFS(2)}(M,N) = \sqrt{\frac{1}{2}((1-0)^2 + (0-1)^2)} = 1 \qquad (3.162)$$

$$e_{IFS(2)}(M,G) = \sqrt{\frac{1}{2}((1-\frac{1}{2})^2 + (0-\frac{1}{2})^2)} = \frac{1}{2} \qquad (3.163)$$

$$e_{IFS(2)}(N,G) = \sqrt{\frac{1}{2}((0-\frac{1}{2})^2 + (1-\frac{1}{2})^2)} = \frac{1}{2} \qquad (3.164)$$

$$e_{IFS(2)}(E,G) = \sqrt{\frac{1}{2}((\frac{1}{4}-\frac{1}{2})^2 + (\frac{1}{4}-\frac{1}{2})^2)} = \frac{1}{4} \qquad (3.165)$$

$$e_{IFS(2)}(H,G) = \sqrt{\frac{1}{2}((0-\frac{1}{2})^2 + (0-\frac{1}{2})^2)} = \frac{1}{4} \qquad (3.166)$$

However, one can hardly agree with the above results. As it was shown (cf. Figure 2.3), the triangle *MNH* (Figure 3.2) has all edges equal to $\sqrt{2}$ (as they are diagonals of squares with sides equal to 1). So we should obtain $e_{IFS(2)}(M,H) = e_{IFS(2)}(N,H) = e_{IFS(2)}(M,N)$. But our results show only that $e_{IFS(2)}(M,H) = e_{IFS(2)}(N,H)$ [cf. (3.160)–(3.161)], but unfortunately $e_{IFS(2)}(M,H) \neq e_{IFS(2)}(M,N)$, and $e_{IFS(2)}(N,H) \neq e_{IFS(2)}(M,N)$. Also $e_{IFS(2)}(E,G)$, which is half of the height of triangle *MNH* multiplied in (3.159) by $\sqrt{1/2}$, is not the value we expect (it is too short, and the same concerns the height of $e_{IFS(2)}(H,G)$).

On the other hand, upon calculating the same Euclidean distances using (3.155), i.e., taking into account all three terms (membership values, non-membership values, and hesitation margins), we obtain:

$$e_{IFS}^1(M,H) = \sqrt{\frac{1}{2}((1-0)^2 + 0^2 + (0-1)^2)} = 1 \qquad (3.167)$$

$$e^1_{IFS}(N,H) = \sqrt{\frac{1}{2}(0^2 + (1-0)^2 + (0-1)^2)} = 1 \qquad (3.168)$$

$$e^1_{IFS}(M,N) = \sqrt{\frac{1}{2}((1-0)^2 + (0-1)^2 + 0^2)} = 1 \qquad (3.169)$$

$$e^1_{IFS}(M,G) = \sqrt{\frac{1}{2}((1-\frac{1}{2})^2 + (0-\frac{1}{2})^2 + 0^2)} = \frac{1}{2} \qquad (3.170)$$

$$e^1_{IFS}(N,G) = \sqrt{\frac{1}{2}((0-\frac{1}{2})^2 + (1-\frac{1}{2})^2 + 0^2)} = \frac{1}{2} \qquad (3.171)$$

$$e^1_{IFS}(E,G) = \sqrt{\frac{1}{2}((\frac{1}{4}-\frac{1}{2})^2 + (\frac{1}{4}-\frac{1}{2})^2 + (\frac{1}{2}-0)^2)} = \frac{\sqrt{3}}{4} \qquad (3.172)$$

$$e^1_{IFS}(H,G) = \sqrt{\frac{1}{2}((0-\frac{1}{2})^2 + (0-\frac{1}{2})^2 + (1-0)^2)} = \frac{\sqrt{3}}{2} \qquad (3.173)$$

From (3.155) we get the expected results, i.e.

$$e^1_{IFS}(M,H) = e^1_{IFS}(N,H) = e^1_{IFS}(M,N) = 2e^1_{IFS}(M,G) = 2e^1_{IFS}(N,G)$$

and $e^1_{IFS}(E,G)$ is equal to half the height of a triangle with all edges equal $\sqrt{2}$ multiplied by $1/\sqrt{2}$, i.e. $\frac{\sqrt{3}}{4}$.

Example 3.4. (Szmidt and Kacprzyk [171]) Let A and B in $X = \{1,2,3,4,5,6,7\}$ be the following intuitionistic fuzzy sets

$$A = (0.5, 0.3, 0.2)/1 + (0.2, 0.6, 0.2)/2 + (0.3, 0.2, 0.5)/4 +$$
$$+ (0.2, 0.2, 0.6)/5 + (1, 0, 0)/6 \qquad (3.174)$$

$$B = (0.2, 0.6, 0.2)/1 + (0.3, 0.2, 0.5)/4 + (0.5, 0.2, 0.3)/5 + (0.9, 0, 0.1)/7 \qquad (3.175)$$

Then, upon taking into account all three terms, we get the Hamming distance (3.154) equal to

$$d^1_{IFS}(A,B) = \frac{1}{2}(|0.5 - 0.2| + |0.3 - 0.6| + |0.2 - 0.2| + |0.2 - 0| + |0.6 - 1| +$$
$$+ |0.2 - 0| + |0.3 - 0.3| + |0.2 - 0.2| + |0.5 - 0.5| + |0.2 - 0.5| +$$
$$+ |0.2 - 0.2| + |0.6 - 0.3| + |1 - 0| + |0 - 1| + |0 - 0| +$$
$$+ |0 - 0.9| + |1 - 0| + |0 - 0.1|) = 3 \qquad (3.176)$$

Thus, taking into account all three terms, we get the normalized Hamming distance (3.156) as equal to

$$l_{IFS}^1(A,B) = \frac{3}{7} = 0.43 \tag{3.177}$$

The Hamming distance, when we account for two terms only, is equal to

$$d_{IFS(2)}(A,B) = \frac{1}{2}(|0.5 - 0.2| + |0.3 - 0.6| + |0.2 - 0| + |0.6 - 1| + |0.3 - 0.3| +$$
$$+ |0.2 - 0.2| + |0.2 - 0.5| + |0.2 - 0.2| + |1 - 0| + |0 - 1| +$$
$$+ |0 - 0.9| + |1 - 0|) = 2.7 \tag{3.178}$$

and the normalized Hamming distance accounting for two terms only is

$$l_{IFS(2)}(A,B) = \frac{1}{7} \cdot d(A,B) = \frac{2.7}{7} = 0.39 \tag{3.179}$$

The Euclidean distance (3.155) based on all three terms is equal to

$$e_{IFS}^1(A,B) = 0.5^{0.5}((0.5 - 0.2)^2 + (0.3 - 0.6)^2 + (0.2 - 0.2)^2 + (0.2 - 0)^2 +$$
$$+ (0.6 - 1)^2 + (0.2 - 0)^2 + (0.3 - 0.3)^2 + (0.2 - 0.2)^2 +$$
$$+ (0.5 - 0.5)^2 + (0.2 - 0.5)^2 + (0.2 - 0.2)^2 + (0.6 - 0.3)^2 +$$
$$+ (1 - 0)^2 + (0 - 1)^2 + 0^2 + (0 - 0.9)^2 + (1 - 0)^2 + (0 - 0.1)^2)^{0.5} =$$
$$= \sqrt{2.21} = 1.49 \tag{3.180}$$

thus, the normalized Euclidean distance based on all three terms is

$$q_{IFS}^1(A,B) = \frac{e_{IFS}^1(A,B)}{\sqrt{7}} = \sqrt{\frac{2.21}{7}} = 0.56 \tag{3.181}$$

The Euclidean distance (3.159), calculated with two terms only is equal to

$$e_{IFS(2)}(A,B) = 0.5^{0.5}((0.5 - 0.2)^2 + (0.3 - 0.6)^2 + (0.2 - 0)^2 + (0.6 - 1)^2 +$$
$$+ (0.3 - 0.3)^2 + (0.2 - 0.2)^2 + (0.2 - 0.5)^2 + (0.2 - 0.2)^2 + (1 - 0)^2 +$$
$$+ (0 - 1)^2 + (0 - 0.9)^2 - (1 - 0)^2)^{0.5} = \sqrt{2.14} = 1.46 \tag{3.182}$$

hence, the normalized Euclidean distance, based on only two terms is

$$q_{IFS(2)}(A,B) = \sqrt{\frac{1}{7} \cdot e(A,B)} = \sqrt{\frac{2.14}{7}} = 0.55 \tag{3.183}$$

It is easy to notice, when analyzing the results obtained in Examples 3.3 and 3.4 that distances between the intuitionistic fuzzy sets should be calculated by taking into account all three terms (membership values, non-membership values, and hesitancy margin values). It is also easy to notice that for the formulas (3.154) –(3.157) the following holds

$$0 \leq d^1_{IFS}(A,B) \leq n \tag{3.184}$$

$$0 \leq l^1_{IFS}(A,B) \leq 1 \tag{3.185}$$

$$0 \leq e^1_{IFS}(A,B) \leq \sqrt{n} \tag{3.186}$$

$$0 \leq q^1_{IFS}(A,B) \leq 1 \tag{3.187}$$

Using two terms only gives values of distances which are orthogonal projections of the real distances (Figure 2.3), and this implies that they are lower.

So to sum up, after analyzing several definitions of distances between the intuitionistic fuzzy sets, it was shown that the distances should be calculated taking into account all three terms describing an intuitionistic fuzzy set.

Taking into account all three terms describing the intuitionistic fuzzy sets when calculating distances ensures that the distances for fuzzy sets and intuitionistic fuzzy sets can be easily compared [cf. formulas (3.140)-(3.143) and formulas (3.184)-(3.187)].

3.3.2.1   Hausdorff Distances

The Hausdorff distances (cf. Grünbaum [77]) are important from the point of view of practical applications, namely, in image matching, image analysis, visual navigation of robots, motion tracking, computer-assisted surgery and so on (cf. e.g., Huttenlocher et al. [89], Huttenlocher and Rucklidge [90], Olson [127], Peitgen et al. [136], Rucklidge [142]-[146]). Although the definition of the Hausdorff distances is simple, the calculations needed to solve the real problems are complex. In result the efficiency of the algorithms for computing the Hausdorff distances may be crucial and the use of some approximations may be relevant and useful (e.g, Aichholzer [1], Atallah [3], Huttenlocher et al. [89], Preparata and Shamos [137], Rucklidge [146], Veltkamp [239]).

First of all, the formulas proposed for calculating the distances should be formally correct. This is the motivation of this section. Namely, we consider the results of using the Hamming distances between the intuitionistic fuzzy sets calculated in two possible ways – taking into account the two term representation (the membership and non-membership values) of the intuitionistic fuzzy sets, and next – taking into account the three term representation (the membership, non-membership values, and hesitation margin values) of the intuitionistic fuzzy sets. We will verify if the resulting distances fulfill the properties of the Hausdorff distances.

The next problem we consider concerns calculating the Hausdorff distance based on the Hamming metric for the interval-valued fuzzy sets. We prove that the formulas that are effective and efficient for the interval-valued fuzzy sets do not work well in the case of the intuitionistic fuzzy sets.

The Hausdorff distance is *the maximum distance of a set to the nearest point in the other set* (Rote [141]). More formal description is given by the following

Definition 3.13. Given two finite sets $A = \{a_1, ..., a_p\}$ and $B = \{b_1, ..., b_q\}$, the Hausdorff distance $H(A, B)$ is defined as:

$$H(A, B) = \max\{h(A, B), h(B, A)\} \qquad (3.188)$$

where

$$h(A, B) = \max_{a \in A} \min_{b \in B} d(a, b) \qquad (3.189)$$

where:

– a and b are elements belonging to sets A and B respectively,
– $d(a, b)$ is any metric between elements a and b,
– the two distances $h(A, B)$ and $h(B, A)$ (3.189) are called the directed Hausdorff distances.

The directed Hausdorff distance from A to B, i.e., the function $h(A, B)$ ranks each element of A based on its distance to the nearest element of B, and then the highest ranked element specifies the value of the distance. Usually, $h(A, B)$ and $h(B, A)$ can be different values (the directed distances are not symmetric).

Following the way of calculating the Hausdorff distances (Definition 3.13) we may notice that if A and B contain one element each (a_1 and b_1, respectively), the Hausdorff distance is just equal to $d(a_1, b_1)$. In other words, if for separate elements a formula which is expected to express the Hausdorff distance gives a result which is not consistent with the used metric d (e.g., the Hamming distance, the Euclidean distance, etc.), the formula considered is not a proper definition of the Hausdorff distance.

3.3.2.2   The Hausdorff Distance Between the InterVal-valued Fuzzy Sets

The Hausdorff distance between two intervals: $U = [u_1, u_2]$ and $W = [w_1, w_2]$ is (Moore [125]):

$$h(U, W) = \max\{|u_1 - w_1|, |u_2 - w_2|\} \qquad (3.190)$$

Assuming the two-term representation for the intuitionistic fuzzy sets: $A = \{x, \mu_A(x), v_A(x)\}$ and $B = \{x, \mu_B(x), v_B(x)\}$, we may consider the two intuitionistic fuzzy sets, A and B, as two intervals, namely:

$$[\mu_A(x), 1 - v_A(x)] \text{ and } [\mu_B(x), 1 - v_B(x)] \qquad (3.191)$$

then

$$h(A, B) = \max\{|\mu_A(x) - \mu_B(x)|, |v_A(x) - v_B(x)|\} \qquad (3.192)$$

Later on we will verify if (3.192) is a properly calculated Hausdorff distance between the intuitionistic fuzzy sets while using the Hamming metric.

3.3.2.3   Two Term Representation of the Intuitionistic Fuzzy Sets and the Hausdorff Distance (Hamming Metric)

Following the algorithm of calculating the directed Hausdorff distances, when applying the two term type Hamming distance (3.88) between the intuitionistic fuzzy sets, we obtain:

$$d_h(A,B) = \frac{1}{n} \sum_{i=1}^{n} max\{|\mu_A(x_i) - \mu_B(x_i)|, |\nu_A(x_i) - \nu_B(x_i)|\} \qquad (3.193)$$

If the above distance (3.193) is a properly calculated Hausdorff distance, then in the case of degenerate, i.e., one-element sets $A = \{< x, \mu_A(x), \nu_A(x) >\}$ and $B = \{< x, \mu_B(x), \nu_B(x) >\}$, it should give the same results as the two term type Hamming distance (3.88). It means that in the case of the two term type Hamming distance, for the degenerate, one element intuitionistic fuzzy sets, the following equations should give just the same results (Szmidt and Kacprzyk [209]):

$$l_{IFS(2)}(A,B) = \frac{1}{2}(|\mu_A(x) - \mu_B(x)| + |\nu_A(x) - \nu_B(x)|) \qquad (3.194)$$

$$d_h(A,B) = max\{|\mu_A(x) - \mu_B(x)|, |\nu_A(x) - \nu_B(x)|\} \qquad (3.195)$$

where (3.194) is the normalized two term type Hamming distance, and (3.195) should be its counterpart Hausdorff distance.

We will verify on a simple example if (3.194) and (3.195) give the same results as they should do following the essence of the Hausdorff measures.

Example 3.5. (Szmidt and Kacprzyk [210]) Consider the following one-element intuitionistic fuzzy sets: $A, B, D, G, E \in X = \{x\}$

$$A = \{< x, 1, 0 >\}, \quad B = \{< x, 0, 1 >\}, \quad D = \{< x, 0, 0 >\},$$
$$G = \{< x, \frac{1}{2}, \frac{1}{2} >\}, \quad E = \{< x, \frac{1}{4}, \frac{1}{4} >\} \qquad (3.196)$$

The results from (3.195) are:

$$d_h(A,B) = max\{|1-0|, |0-1|\} = 1$$

$$d_h(A,D) = max\{|1-0|, |0-0|\} = 1$$

$$d_h(B,D) = max\{|0-0|, |1-0|\} = 1$$

$$d_h(A,G) = max\{|1-1/2|, |0-1/2|\} = 0.5$$

$$d_h(A,E) = max\{|1-1/4|, |0-1/4|\} = 0.75$$

$$d_h(B,G) = \max\{|0 - 1/2|, |1 - 1/2|\} = 0.5$$

$$d_h(B,E) = \max\{|0 - 1/4|, |1 - 1/4|\} = 0.75$$

$$d_h(D,G) = \max\{|0 - 1/2|, |0 - 1/2|\} = 0.5$$

$$d_h(D,E) = \max\{|0 - 1/4|, |1 - 1/4|\} = 0.25$$

$$d_h(G,E) = \max\{|1/2 - 1/4|, |1/2 - 1/4|\} = 0.25$$

Their counterpart Hamming distances calculated from (3.194) are:

$$l_{IFS(2)}(A,B) = 0.5(|1 - 0| + |0 - 1|) = 1$$

$$l_{IFS(2)}(A,D) = 0.5(|1 - 0| + |0 - 0||) = 0.5$$

$$l_{IFS(2)}(B,D) = 0.5(|0 - 0| + |1 - 0||) = 0.5$$

$$l_{IFS(2)}(A,G) = 0.5(|0 - 1/2| + |0 - 1/2|) = 0.5$$

$$l_{IFS(2)}(A,E) = 0.5(|1 - 1/4| + |0 - 1/4||) = 0.5$$

$$l_{IFS(2)}(B,G) = 0.5(|1 - 1/4| + |0 - 1/4|) = 0.5$$

$$l_{IFS(2)}(B,E) = 0.5(|1 - 1/4| + |0 - 1/4|) = 0.5$$

$$l_{IFS(2)}(D,G) = 0.5(|0 - 1/2| + |0 - 1/2|) = 0.5$$

$$l_{IFS(2)}(D,E) = 0.5(|0 - 1/4| + |0 - 1/4|) = 0.25$$

$$l_{IFS(2)}(G,E) = 0.5(|1/2 - 1/4| + |1/2 - 1/4|) = 0.25$$

i.e. the values of the Hamming distances (3.194) used to define the Hausdorff measures (3.195), and the values of the resulting Hausdorff distances (3.195) calculated for the separate elements are not consistent (as they should be). The differences are:

$$d_h(A,D) \neq l_{IFS(2)}(A,D) \qquad\qquad (3.197)$$

$$d_h(B,D) \neq l_{IFS(2)}(B,D) \qquad (3.198)$$

$$d_h(A,E) \neq l_{IFS(2)}(A,E) \qquad (3.199)$$

$$d_h(B,E) \neq l_{IFS(2)}(B,E) \qquad (3.200)$$

It is easy to show that the inconsistencies as shown above occur for an infinite number of other cases.

Now we will verify the conditions under which the equations (3.194) and (3.195) give consistent results, i.e., when for the separate elements we have (Szmidt and Kacprzyk [218]):

$$\frac{1}{2}(|\mu_A(x) - \mu_B(x)| + |\nu_A(x) - \nu_B(x)|) =$$
$$= \max\{|\mu_A(x) - \mu_B(x)|, |\nu_A(x) - \nu_B(x)|\} \qquad (3.201)$$

Taking into account that

$$\mu_A(x) + \nu_A(x) + \pi_A(x) = 1 \qquad (3.202)$$

$$\mu_B(x) + \nu_B(x) + \pi_B(x) = 1 \qquad (3.203)$$

from (3.202) and (3.203) we obtain

$$(\mu_A(x) - \mu_B(x)) + (\nu_A(x) - \nu_B(x)) + (\pi_A(x) - \pi_B(x)) = 0 \qquad (3.204)$$

It is easy to notice that (3.204) is not fulfilled for all elements belonging to an intuitionistic fuzzy set but for some elements only. Namely, equation (3.201) is fulfilled for the following conditions (Szmidt and Kacprzyk [218])

- for $\pi_A(x) - \pi_B(x) = 0$, from (3.204) we have

$$|\mu_A(x) - \mu_B(x)| = |\nu_A(x) - \nu_B(x)| \qquad (3.205)$$

and having in mind (3.205), we can express (3.201) in the following way:

$$0.5(|\mu_A(x) - \mu_B(x)| + |\mu_A(x) - \mu_B(x)|) =$$
$$= \max\{|\mu_A(x) - \mu_B(x)|, |\mu_A(x) - \mu_B(x)|\} \qquad (3.206)$$

- if $\pi_A(x) - \pi_B(x) \neq 0$, but, at the same time

$$\mu_A(x) - \mu_B(x) = \nu_A(x) - \nu_B(x) = -\frac{1}{2}(\pi_A(x) - \pi_B(x)) \qquad (3.207)$$

then (3.201) boils down again to (3.206).

In other words, (3.201) is fulfilled (which means that the Hausdorff measure given by (3.195) is a natural counterpart of (3.194)) only for such elements belonging to an intuitionistic fuzzy set, for which some additional conditions are given, like $\pi_A(x) - \pi_B(x) = 0$ or (3.207). However in general, for an infinite numbers of elements, (3.201) is not valid.

In the above context it seems to be a bad idea to try constructing the Hausdorff distance using the two term type Hamming distance between the intuitionistic fuzzy sets.

An immediate conclusion is that, relating to the results concerning interval-valued fuzzy sets (3.190)–(3.192) the Hausdorff distance for the intuitionistic fuzzy sets can not be constructed in the same way as for the interval-valued fuzzy sets.

3.3.2.4 Three Term Hamming Distance Between the Intuitionistic Fuzzy Sets and the Hausdorff Metric

Now we will show that by applying the three term type Hamming distance for the intuitionistic fuzzy sets, we obtain correct (in the sense of Definition 3.13) Hausdorff distance.

Namely, if we calculate the three term type Hamming distance between two degenerate, i.e. one-element intuitionistic fuzzy sets, A and B in the spirit of Szmidt and Kacprzyk [171], [188], Szmidt and Baldwin [159], [160], i.e., in the following way:

$$l^1_{IFS}(A,B) = \frac{1}{2}(|\mu_A(x) - \mu_B(x)| + |\nu_A(x) - \nu_B(x)| + \\ + |\pi_A(x) - \pi_B(x)|) \tag{3.208}$$

we can give a counterpart of the above distance in terms of the max function (Szmidt and Kacprzyk [218]):

$$H_3(A,B) = \max\{|\mu_A(x) - \mu_B(x)|, |\nu_A(x) - \nu_B(x)|, \\ , |\pi_A(x) - \pi_B(x)|\} \tag{3.209}$$

If $H_3(A,B)$ (3.209) is a properly specified Hausdorff distance (in the sense that for two degenerate, one element intuitionistic fuzzy sets, the result is equal to the metric used), the following condition should be fulfilled (Szmidt and Kacprzyk [218]):

$$\frac{1}{2}(|\mu_A(x) - \mu_B(x)| + |\nu_A(x) - \nu_B(x)|) + |\pi_A(x) - \pi_B(x)|) = \\ = \max\{|\mu_A(x) - \mu_B(x)|, |\nu_A(x) - \nu_B(x)|, |\pi_A(x) - \pi_B(x)|\} \tag{3.210}$$

Let us verify if (3.210) is valid. Without loss of generality we can assume

$$\max\{|\mu_A(x) - \mu_B(x)|, |\nu_A(x) - \nu_B(x)|, |\pi_A(x) - \pi_B(x)|\} = \\ = |\mu_A(x) - \mu_B(x)| \tag{3.211}$$

For $|\mu_A(x) - \mu_B(x)|$ fulfilling (3.211), and because of (3.202) and (3.203), we conclude that both $\nu_A(x) - \nu_B(x)$, and $\pi_A(x) - \pi_B(x)$ are of the same sign (both values are either positive or negative). Therefore

$$|\mu_A(x) - \mu_B(x)| = |\nu_A(x) - \nu_B(x)| + |\pi_A(x) - \pi_B(x)| \qquad (3.212)$$

Applying (3.212) we can verify that (3.210) always is valid as

$$0.5\{|\mu_A(x) - \mu_B(x)| + |\mu_A(x) - \mu_B(x)|\} =$$
$$= \max\{|\mu_A(x) - \mu_B(x)|, |\nu_A(x) - \nu_B(x)|, |\pi_A(x) - \pi_B(x)|\} =$$
$$= |\mu_A(x) - \mu_B(x)| \qquad (3.213)$$

Now we will use the above formulas, (3.208) and (3.209), for the data used in Example 1. But now, as we also take into account the hesitation margins $\pi(x)$ (2.7), instead of (3.196) we use the three term, "full" description of the data $\{< x, \mu(x), \nu(x), \pi(x) >\}$, i.e. employing all three functions (the membership, non-membership and hesitation margin) describing the considered intuitionistic fuzzy sets (Szmidt and Kacprzyk [210]):

$$A = \{< x, 1, 0, 0 >\}, \quad B = \{< x, 0, 1, 0 >\}, \quad D = \{< x, 0, 0, 1 >\},$$
$$G = \{< x, \frac{1}{2}, \frac{1}{2}, 0 >\}, \quad E = \{< x, \frac{1}{4}, \frac{1}{4}, \frac{1}{2} >\} \qquad (3.214)$$

From (3.209) we have:

$$H_3(A,B) = \max(|1-0|, |0-1|, |0-0|) = 1$$

$$H_3(A,D) = \max(|1-0|, |0-0|, |0-1|) = 1$$

$$H_3(B,D) = \max(|0-0|, |1-0|, |0-1|) = 1$$

$$H_3(A,G) = \max(|0-1/2|, |0-1/2|, |0-0|) = 0.5$$

$$H_3(A,E) = \max(|1-1/4|, |0-1/4|, |0-1/2|) = 0.75$$

$$H_3(B,G) = \max(|1-1/4|, |0-1/4|, |0-1/2|) = 0.75$$

$$H_3(B,E) = \max(|1-1/4|, |0-1/4|, |0-1/2|) = 0.75$$

$$H_3(D,G) = \max(|0-1/2|, |0-1/2|, |1-0|) = 1$$

$$H_3(D,E) = \max(|0-1/4|, |0-1/4|, |1-1/2|) = 0.5$$

$$H_3(G,E) = \max(|1/2 - 1/4|, |1/2 - 1/4|, |0 - 1/2|) = 0.5$$

The counterpart Hamming distances obtained from (3.208) (with all three functions) are

$$l^1_{IFS}(A,B) = 0.5(|1 - 0| + |0 - 1| + |0 - 0|) = 1$$

$$l^1_{IFS}(A,D) = 0.5(|1 - 0| + |0 - 0| + |0 - 1|) = 1$$

$$l^1_{IFS}(B,D) = 0.5(|0 - 0| + |1 - 0| + |0 - 1|) = 1$$

$$l^1_{IFS}(A,G) = 0.5(|0 - 1/2| + |0 - 1/2| + |0 - 0|) = 0.5$$

$$l^1_{IFS}(A,E) = 0.5(|1 - 1/4| + |0 - 1/4| + |0 - 1/2|) = 0.75$$

$$l^1_{IFS}(B,G) = 0.5(|1 - 1/4| + |0 - 1/4| + |0 - 1/2|) = 0.75$$

$$l^1_{IFS}(B,E) = 0.5(|1 - 1/4| + |0 - 1/4| + |0 - 1/2|) = 0.75$$

$$l^1_{IFS}(D,G) = 0.5(|0 - 1/2| + |0 - 1/2| + |1 - 0|) = 1$$

$$l^1_{IFS}(D,E) = 0.5(|0 - 1/4| + |0 - 1/4| + |1 - 1/2|) = 0.5$$

$$l^1_{IFS}(G,E) = 0.5(|1/2 - 1/4| + |1/2 - 1/4| + |0 - 1/2|) = 0.5$$

As we can see, the Hausdorff distance (3.209) (using the membership values, non-membership values and hesitation margins) and the tree term Hamming distance (3.208) give for one-element intuitionistic fuzzy sets fully consistent results. The same situation occurs in a general case too.

In other words, for the normalized Hamming distance expressed in the spirit of (Szmidt and Kacprzyk [171], [188]), given by (3.154), we can give the following equivalent representation in terms of the max function:

$$H_3(A,B) = \frac{1}{n} \sum_{i=1}^{n} \max \{ |\mu_A(x_i) - \mu_B(x_i)|, |\nu_A(x_i) - \nu_B(x_i)|,$$
$$|\pi_A(x_i) - \pi_B(x_i)| \} \qquad (3.215)$$

Unfortunately, it can be easily verified that it is impossible to give the counterpart pairs of the formulas like (3.154) and (3.215) for $r > 1$ in the Minkowski r-metrics

($r = 1$ is the Hamming distance, $r = 2$ is the Euclidean distance, etc.). More details are given in [25] and [236].

Now we will show that the three term distances between the intuitionistic fuzzy sets are useful in the ranking of intuitionistic fuzzy alternatives.

3.4 Ranking of the Intuitionistic Fuzzy Alternatives

Given their ability to model imperfect information, the intuitionistic fuzzy sets have found applications in many areas, in particular, in decision making. Ranking of the intuitionistic fuzzy alternatives (options), obtained, for example, as a result of deci- sion analysis, aggregation, etc. is one of important problems. The intuitionistic fuzzy alternatives may be understood in different ways. Here we mean them as elements of a universe of discourse with their associated membership degrees, non-membership degrees, and hesitation margins. In the context of decision making each option ful- fills a set of criteria to some extent $\mu(.)$, it does not fulfill this set of criteria to some extent $\nu(.)$ and, on the other hand we are not sure to the extent $\pi(.)$ if an option fulfills or does not fulfill a set of criteria. This implies that the alternatives can be expressed via the intuitionistic fuzzy sets. Here we will call such alternatives "intuitionistic fuzzy alternatives".

The intuitionistic fuzzy alternatives may be ranked only under some additional assumptions as there is no linear order among elements of the intuitionistic fuzzy sets. The situation is different from that for fuzzy sets (Zadeh [254]), for which ele- ments of the universe of discourse are naturally ordered because their membership degrees are real numbers from $[0, 1]$.

There are not many approaches for ranking the intuitionistic fuzzy alternatives in the literature. For instance, Chen and Tan [53], Hong and Choi [80], Li et al. [112], [114], and Hua-Wen Liu and Guo-Jun Wang [119] proposed some approaches.

Chen and Tan [53] proposed a score function for vague sets [72], but, as Bustince and Burillo [44] demonstrated that vague sets are equivalent to intuitionistic fuzzy sets, we can consider the concept of a score function for an intuitionistic fuzzy alternative $a = (\mu, \nu)$ meant as

$$S(a) = \mu - \nu, \qquad (3.216)$$

and, clearly, $S(a) \in [-1, 1]$.

It is easy to notice that the score function $S(a)$ (3.216) can not alone evaluate the intuitionistic fuzzy alternatives as it produces the same result for such different intuitionistic fuzzy alternatives $a = (\mu, \nu)$ as, e.g.: $(0.5, 0.4)$, $(0.4, 0.3)$, $(0.3, 0.2)$, $(0.1, 0)$ – for all of them $S(a) = 0.1$, which seems counterintuitive.

Next, Hong and Choi [80] introduced, in addition to the score function (3.216), a so called accuracy function H

$$H(a) = \mu + \nu, \qquad (3.217)$$

where $H(a) \in [0, 1]$.

Xu [253] made use of both (3.216) and (3.217), and proposed an algorithm rank-ing the intuitionistic fuzzy alternatives. In the case of two alternatives a_i and a_j, the algorithm is as follows [253]:

- if $S(a_i) \leq S(a_j)$, then a_i is smaller than a_j;
- if $S(a_i) = S(a_j)$, then:

  - if $H(a_i) = H(a_j)$, then a_i and a_j represent the same information (are equal);
  - if $H(a_i) \leq H(a_j)$, then a_i is smaller than a_j.

Unfortunately, the above method of ranking does not produce reliable results in many cases. Let us consider two intuitionistic fuzzy alternatives (Szmidt and Kac-przyk [205]) $a_1 - (0.5, 0.45)$ and $a_2 - (0.25, 0.05)$ for which we obtain $S(a_1) = 0.5 - 0.45 = 0.05$, $S(a_2) = 0.25 - 0.05 = 0.2$, suggesting that a_1 is smaller than a_2. However, information provided by a_1 (i.e. $0.5 + 0.45 = 0.95$) is certainly bigger than that provided by a_2 (i.e. $0.25 + 0.05 = 0.3$). In this context it is difficult to agree that a_1 is smaller than a_2. Later on, we will return to ranking of two intuitionistic fuzzy alternatives by the method we propose.

We give below an example showing some more weak sides of the above proce-dure. Let us consider the following intuitionistic fuzzy alternatives:

$$a_1 = (0.1, 0, 0.9),$$
$$a_2 = (0.2, 0.11, 0.69),$$
$$a_3 = (0.3, 0.22, 0.48),$$
$$a_4 = (0.4, 0.33, 0.27),$$
$$a_5 = (0.5, 0.44, 0.06),$$

for which the scores are:

$$S(a_1) = 0.1 - 0 = 0.1,$$
$$S(a_2) = 0.2 - 0.11 = 0.09,$$
$$S(a_3) = 0.3 - 0.22 = 0.08,$$
$$S(a_4) = 0.4 - 0.33 = 0.07,$$
$$S(a_5) = 0.5 - 0.44 = 0.06,$$

which, in the light of the above algorithm means that:

$$a_1 > a_2 > a_3 > a_4 > a_5$$

In other words, due to the above ranking procedure, in this particular case, the less we know, the better (it is worth noticing that the lack of knowledge is the biggest for the "best" alternative a_1 (equal to 0.9), and it decreases for the consecutive "worse" (according to the considered procedure) alternatives. Next, the membership values increase from 0.1 (for a_1) to 0.5 (for a_5). Thus, for increasing membership values and decreasing lack of knowledge we obtain (from the ranking procedure consid-ered) worse alternatives, which is obviously counterintuitive.

Moreover, the ranking procedure considered produces answers that are not continuous. If we change a little the non-membership values in the above example, i.e.:

$$a_1 = (0.1, 0, 0.9),$$
$$a_2 = (0.2, 0.1, 0.7),$$
$$a_3 = (0.3, 0.2, 0.5),$$
$$a_4 = (0.4, 0.3, 0.3),$$
$$a_5 = (0.5, 0.4, 0.1),$$

we obtain the same score for each $a_i, i = 1, \ldots, 5$, and from the second part of the ranking procedure we obtain the reverse order, i.e.:

$$a_5 > a_4 > a_3 > a_2 > a_1$$

Certainly, it makes no sense for a ranking procedure to be so sensitive to so small changes of the parameters. Conclusion: the above ranking procedure should not be used (especially in decision making tasks).

We have already mentioned the possibility of using the intuitionistic fuzzy sets in voting models. Now we will consider some ways of ranking the voting alternatives expressed via the intuitionistic fuzzy elements.

Let an element x belonging to an intuitionistic fuzzy set characterized via (μ, ν, π) express a voting situation: μ represents the proportion (from $[0, 1]$) of voters who vote for x, ν represents the proportion of those who vote against x, and π represents the proportion of those who abstain. The simplest idea of comparing different voting situations (ranking the alternatives) would be to use a distance measure from the ideal voting situation $M = (x, 1, 0, 0)$ (100% voting for, 0% vote against and 0% abstain) to the alternatives considered. We will call M the ideal positive alternative. Let

$A = (x, 0.5, 0.5, 0)$ – 50% vote for, 50% against, and 0% abstain,
$B = (x, 0.4, 0.4, 0.2)$ – 40% vote for, 40% vote against and 20% abstain,
$C = (x, 0.3, 0.3, 0.4)$ – 30% vote for, 30% vote against and 40% abstain.

First we confirm that the method of calculating distances between two intuitionistic fuzzy sets A and B described by two terms, i.e., the membership and non-membership values only (3.218) does not work properly (cf. Szmidt and Kacprzyk [171], [188], Szmidt and Baldwin [159], [160]) in this case, too:

$$l_{IFS(2)}(A, B) = \frac{1}{2n} \sum_{i=1}^{n} (|\mu_A(x_i) - \mu_B(x_i)| + |\nu_A(x_i) - \nu_B(x_i)|) \qquad (3.218)$$

The results obtained with (3.218), i.e., the distances for the above voting alternatives represented by points A, B, C (cf. Figure 3.3) from the ideal positive alternative represented by $M(1, 0, 0)$ are, respectively (Szmidt and Kacprzyk [197]):

$$l_{IFS(2)}(M, A) = 0.5(|1 - 0.5| + |0 - 0.5|) = 0.5 \qquad (3.219)$$

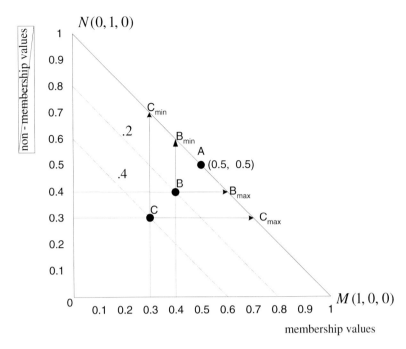

Fig. 3.3 Geometrical representation of the intuitionistic fuzzy alternatives

$$l_{IFS(2)}(M,B) = 0.5(|1 - 0.4| + |0 - 0.4|) = 0.5 \qquad (3.220)$$
$$l_{IFS(2)}(M,C) = 0.5(|1 - 0.3| + |0 - 0.3|) = 0.5 \qquad (3.221)$$

The results seem to be counterintuitive as (3.218) suggests that all the alternatives (represented by) A, B, C are "the same". On the other hand, the normalized Hamming distance (3.156), taking into account, besides the membership and non-membership, also the hesitation margin, gives:

$$l^1_{IFS}(M,A) = 0.5(|1 - 0.5| + |0 - 0.5| + |0 - 0|) = 0.5 \qquad (3.222)$$
$$l^1_{IFS}(M,B) = 0.5(|1 - 0.4| + |0 - 0.4| + |0 - 0.2|) = 0.6 \qquad (3.223)$$
$$l^1_{IFS}(M,C) = 0.5(|1 - 0.3| + |0 - 0.3| + |0 - 0.4|) = 0.7 \qquad (3.224)$$

It is not difficult to accept the results (3.222)–(3.224), reflecting our intuition. Alternative A (cf. Figure 3.3) seems to be the best in the sense that the distance $l_{IFS}(M,A)$ is the smallest (we know for sure that 50% vote for, 50% vote against). The alternative represented by A is just a fuzzy alternative (A lies on MN where the values of the hesitation margin are equal 0). Alternatives B and C, on the other hand, are "less sure" (with the hesitation margins equal 0.2, and 0.4, respectively).

Unfortunately, a weak point in the ranking of alternatives by calculating the distances from the ideal positive alternative represented by M is that for a fixed membership value, from (3.156) we obtain just the same value (for example, if the membership value μ is equal 0.8, for any intuitionistic fuzzy element, i.e. such that

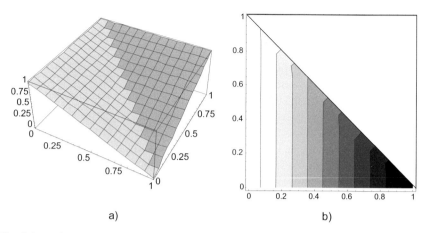

Fig. 3.4 a) Distances (3.156) of any intuitionistic fuzzy element from the ideal alternative M; b) contour plot

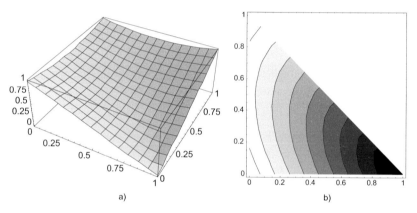

Fig. 3.5 a) Distances (3.157) of any intuitionistic fuzzy element from the ideal alternative M; b) contour plot

its non-membership degree v and hesitation margin π fulfill $v + \pi = 0.2$, we obtain the value of 0.2). This fact is illustrated in Figure 3.4, a and b. To better see this, the distances (3.156) for any alternative from M (Figure 3.4a) are presented for μ and v for the whole range $[0, 1]$ (instead of showing them for $\mu + v \leq 1$ only). For the same reason (to better see the effect), in Figure 3.4b the contour plot of the distances (3.156) is given only for the range of μ and v for which $\mu + v \leq 1$.

Now we will verify if the normalized Euclidean distance (3.157) from the ideal positive alternative represented by $M(1,0,0)$ gives better results from the point of view of ranking the alternatives.

Let $A = (x, 0.2, 0.8, 0)$ – 20% vote for, 80% against, and 0% abstain, $B = (x, 0.2, 0, 0.8)$ – 20% vote for, 0% vote against and 80% abstain, The normalized Euclidean distance (3.157) gives (Szmidt and Kacprzyk [214]) :

$$e^1_{IFS}(M,A) = (0.5((1-0.2)^2 + (0-0.8)^2 + (0-0)^2))^{0.5} = 0.8 \quad (3.225)$$
$$e^1_{IFS}(M,B) = (0.5((1-0.2)^2 + (0-0)^2 + (0-0.8)^2))^{0.5} = 0.8 \quad (3.226)$$

Making use of (3.157) for ranking the alternatives suggests [cf. (3.225)–(3.226)] that the alternatives (represented by) A, B seem to be "the same" which is counterintuitive. A general illustration of the above counterintuitive result is given in Fig. 3.5. We can see that the results of (3.157) are not univocally given for a given membership value μ; for clarity, the distances (3.157) for any x from M (Fig. 3.5a) are presented for μ and v for $[0,1]$, and not for $\mu + v \leq 1$ only. For the same reason (to better see the effect), in Fig. 3.5b the contour plot of the distances (3.157) is given only for the range of μ and v for which $\mu + v < 1$. So, the distances (3.157) (cf. also Szmidt and Kacprzyk [197]) from the ideal positive alternative alone do not make it possible to rank the alternatives in the intended way.

The analysis of the above examples shows that the distances from the ideal positive alternative alone do not make it possible to rank the alternatives in the intended way.

3.4.0.5   A New Method for Ranking Alternatives (Szmidt and Kacprzyk [205])

The sense of a voting alternative (expressed via an intuitionistic fuzzy element) can be analyzed by using the operators (cf. Atanassov [15]) of: *necessity* ($\square$), *possibility* ($\lozenge$), $D_\alpha(A)$ and $F_{\alpha,\beta}(A)$ given as:

- The *necessity* operator ($\square$)

$$\square A = \{\langle x, \mu_A(x), 1 - \mu_A(x)\rangle | x \in X\} \quad (3.227)$$

- The *possibility* operator ($\lozenge$)

$$\lozenge A = \{\langle x, 1 - v_A(x), v_A(x)\rangle | x \in X\} \quad (3.228)$$

- Operator $D_\alpha(A)$ (where $\alpha \in [0,1]$)

$$D_\alpha(A) = \{\langle x, \ \mu_A(x) + \alpha \pi_A(x), \ v_A(x)(1-\alpha)\pi_A(x)\rangle | x \in X\} \quad (3.229)$$

- Operator $F_{\alpha,\beta}(A)$ (where $\alpha, \beta \in [0,1]; \ \alpha + \beta \leq 1$)

$$F_{\alpha,\beta}(A) = \{\langle x, \ \mu_A(x) + \alpha \pi_A(x), v_A(x)\beta \pi_A(x)\rangle | x \in X\} \quad (3.230)$$

Considering alternative $B(0.4, 0.4, 0.2)$, for example, and using the above operators we obtain $\square B = B_{min}$, where $B_{min} = (0.4, 0.6)$, and $\lozenge B = B_{max}$, where $B_{max} = (0.6, 0.4)$ (Figure 3.3). Operator $F_{\alpha,\beta}(A)$ makes it possible for alternative B to become any alternative represented in triangle $BB_{max}B_{min}$. A similar reasoning leads to the conclusion that alternative C (Figure 3.3) might become any

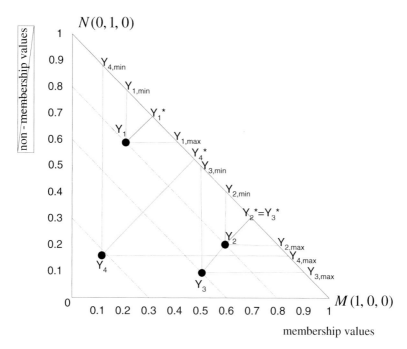

Fig. 3.6 Ranking of alternatives Y_i

alternative represented in triangle $CC_{max}C_{min}$, and alternative $O(0,0,1)$ (with the hesitation margin equal 1)

may become any alternative (the whole area of the triangle MNO).

In the context of the above considerations we could say that the smaller the area of the triangle $Y_iY_{i,min}Y_{i,max}$ (Figure 3.6) the better the alternative Y_i from a set Y of the alternatives considered. Alternatives having their representations on segment MN (i.e., fuzzy alternatives) are the best in the sense that:

- the alternatives are fully reliable in the sense of the information represented, as the hesitation margin is equal 0 here, and
- the alternatives are ordered – the closer an alternative to ideal positive alternative $M(1,0,0)$, the better it is (it is an obvious fact as fuzzy alternatives are univocally ordered).

The above reasoning suggests that a promising way of ranking the intuitionistic fuzzy alternatives Y_i with the same values of π_i is to convert them into the fuzzy alternatives (which may be easily ranked).

The simplest way of ranking the alternatives Y_i with different values of π_i seems to be to make use of the information carried by the triangles $Y_iY_{i,min}Y_{i,max}$.

The amount of information connected with Y_i is indicated by Y_i^*, i.e., by "the position" of triangle $Y_iY_{i,min}Y_{i,max}$ inside triangle MNO – expressed by the projection

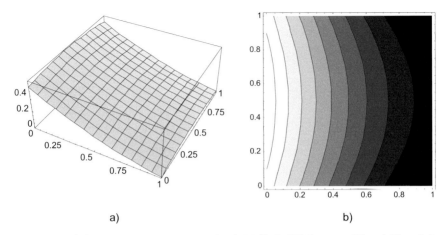

a)   b)

Fig. 3.7 a) $R(Y_i)$ as a function of distance $l_{IFS}(M, Y_i^*)$ (3.156) between Y_i^* and M, and the hesitation margin; b) contour plot

on the segment MN. The hesitation margin π_{Y_i} indicates how reliable (sure) is the information represented by Y_i^*.

Y_i^* are the orthogonal projections of Y_i on MN. Such an orthogonal projection of the intuitionistic fuzzy elements belonging to an intuitionistic fuzzy set A was considered by Szmidt and Kacprzyk [166]. This orthogonal projection may be obtained via operator $D_\alpha(A)$ (3.229) with parameter α equal 0.5.

We can see that all the elements from the segment OA (Figure 3.3) are transformed by $D_{0.5}(A)$ (3.229) into $A(0.5, 0.5)$ which reflects the lack of differences between the membership and non-membership, irrespective of the value of the hesitation margin.

Having the above observations in mind, a reasonable measure R that can be used for ranking the alternatives (represented by) Y_i seems to be

$$R(Y_i) = 0.5(1 + \pi_{Y_i}) distance(M, Y_i^*) \qquad (3.231)$$

where $distance(M, Y_i^*)$ is a distance from the ideal positive alternative $M(1, 0, 0)$, Y_i^* is the orthogonal projection of Y_i on MN. The constant 0.5 was introduced in (3.231) to ensure that $0 < R(Y_i) \leq 1$. The values of function R for any intuitionistic fuzzy element and the distance $l_{IFS}(M, Y_i^*)$ (3.156) are presented in Figure 3.7a, and the corresponding contour plot – in Figure 3.7b.

Unfortunately, the results obtained with (3.231) do not rank the alternatives in the intended way. (The maximum value of (3.231) is not obtained for the alternative $(0, 0, 1)$ but for $(0, 1/2, 1/2)$.)

Similarly, in the case of the normalized Euclidean distance (3.157) used in (3.231) instead of $l_{IFS}^1(M, Y_i^*)$ (3.156), the results of (3.231) do not meet our expectations in the sense of their relations to the areas of the triangles $Y_i Y_{i,min} Y_{i,max}$ (cf. Figure 3.6). Let us consider the alternatives Y_i, $i = 1, \ldots, 4$. of Figure 3.6. We might expect that the alternatives be ordered by (3.231) from Y_1 to Y_4 as just such an order renders the areas of the respective triangles. But the results from (3.231) obtained using the normalized Euclidean distance (3.157) for the different alternatives seem to be very much "the same". For example (Szmidt and Kacprzyk [214]), for $Y_1 = (0, 0.8, 0.2)$, $R_E(Y_1^*) = 0.54$, for $Y_2 = (0, 0.6, 0.4)$, $R_E(Y_2^*) = 0.56$, for $Y_3 = (0, 0.3, 0.7)$, $R_E(Y_3^*) = 0.55$, for $Y_4 = (0, 0, 1)$, $R_E(Y_4^*) = 0.5$.

So, again, the results obtained via (3.231) with the normalized Euclidean distance (3.157) do not rank the alternatives in the intended way.

It seems that a better measure than (3.231) for ranking the alternatives (represented by) Y_i might be the following measure R

$$R(Y_i) = 0.5(1 + \pi_{Y_i})distance(M, Y_i) \tag{3.232}$$

where *distance* means a distance (3.156) of Y_i from the ideal positive alternative $M(1,0,0)$.

Definition (3.232) tells us about the "quality" of an alternative – the lower the value of $R(Y_i)$, (3.232), the better the alternative in the sense of the amount of positive information included, and reliability of information.

For the distance $l_{IFS}^1(M, Y_i)$ (3.156), the best is alternative $M(1,0,0)$ for which $R(M) = 0$. For the alternative $N(0,1,0)$ we obtain $R(N) = 0.5$ (alternative N is fully reliable as the hesitation margin is equal 0, but the distance $l_{IFS}^1(M, N) = 1$). The alternative A (Figures 3.3) gives $R(A) = 0.25$. In general, on MN, the values of R decrease from 0.5 (for alternative N) to 0 (for the best alternative M). The maximum value of R, i.e. 1, is obtained for $O(0,0,1)$ for which both distances from M and the hesitation margin are equal 1 (alternative O "indicates" the whole triangle MNO). All other alternatives Y_i "indicate" smaller triangles $Y_i Y_{i,min} Y_{i,max}$ (Figure 3.6), so their corresponding values of R are smaller (better in the sense of amount of reliable information).

The values of function R (3.232) for any intuitionistic fuzzy element are presented in Figure 3.8a, and the counterpart contour plot – in Figure 3.8b. Considering the numbers obtained via R (3.232), we may notice that the value 0.25 obtained for the alternative $(0.5, 0.5, 0)$ constitutes the "border" of the "interesting" alternatives – in the sense of the amount of positive knowledge.

Let us consider again the ranking of two alternatives (which were ranked counterintuitively by the algorithm presented in [253] as shown in the beginning of Section 3.4), namely $Y_1 = (0.5, 0.45, 0.05)$ and $Y_2 = (0.25, 0.05, 0.7)$ (we stress here that we take into account all three terms: the degrees of membership, non-membership and hesitation margin). From (3.232) we obtain: $R(Y_1) = 0.26$, $R(Y_2) = 0.64$ which means that Y_1 is better than Y_2 (previously, according to the algorithm from [253] Y_2 was better/bigger than Y_1). Obviously, Y_1 is not a "good" option as $R(Y_1)$ is bigger than 0.25 which follows from the fact that the non-membership value is quite big (equal 0.45). It might mean that we would not accept the option Y_1. But option Y_2 seems even less interesting – with the smaller membership value (equal 0.25 instead

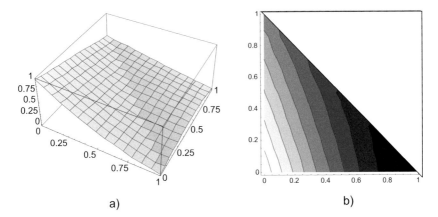

a)   b)

Fig. 3.8 a) $R(Y_i)$ as a function of distance $l_{IFS}(M,Y_i)$ (3.156) between Y_i and M, and the hesitation margin; b) contour plot

of 0.5 for Y_1), and with the bigger hesitation margin (equal 0.7 instead of 0.05 for Y_1).

Example 3.6. (Szmidt and Kacprzyk [205]) Let us evaluate (rank) six medical treatments. The treatments $C1 - C6$, affect a patient in the following way (Szmidt and Kacprzyk [205]):

- $C1 : (0.6, 0.2, 0.2)$ – influences in a positive way 60% of symptoms, in a negative way – 20% of symptoms, and its impact is unknown (was not confirmed) in the case of 20% of symptoms;
- $C2 : (0.7, 0.3, 0)$ – influences in a positive way 70% symptoms, in a negative way – 30% of symptoms, and its impact is unknown (was not confirmed) in the case of 0% of symptoms;
- $C3 : (0.7, 0.15, 0.15)$ – influences in a positive way 70% of symptoms, in a negative way – 15% of symptoms, and its impact is unknown (was not confirmed) in the case of 15% of symptoms;
- $C4 : (0.775, 0.225, 0)$ – influences in a positive way 77.5% of symptoms, in a negative way – 22.5% of symptoms, and its impact is unknown (was not confirmed) in the case of 0% of symptoms;
- $C5 : (0.8, 0.1, 0.1)$ – influences in a positive way 80% of symptoms, in a negative way – 10% of symptoms, and its impact is unknown (was not confirmed) in the case of 10% of symptoms;
- $C6 : (0.8, 0.2, 0)$ – influences in a positive way 80% of symptoms, in a negative way – 20% of symptoms, and its impact is unknown (was not confirmed) in the case of 0% of symptoms.

Table 3.1 shows the ranking of $C1, \ldots, C6$ with (3.232) – from the worst one, $C1$ to the best one, $C6$.

It is worth emphasizing that the ranking function R (3.232) is constructed taking strongly into account the lack of knowledge. Let us consider the pair: $C1$ and $C2$

Table 3.1 Ranking alternatives by R (3.232) – results for the data from Example 3.6

No.	$C_i : (\mu_i, \nu_i, \pi_i)$	$R_E(C_i)$
1	$C1 : (0.6, 0.2, 0.2)$	0.240
2	$C2 : (0.7, 0.3, 0)$	0.150
3	$C3 : (0.7, 0.15, 0.15)$	0.173
4	$C4 : (0.775, 0.225, 0)$	0.113
5	$C5 : (0.8, 0.1, 0.1)$	0.110
6	$C6 : (0.8, 0.2, 0)$	0.100

(Table 3.1). In the case of $C1$ the lack of knowledge is equal to 0.2, so that theoretically, we might expect "on the average" that the hesitation margin representing the lack of knowledge will be divided equally between the membership value and non-membership value giving as a result the case $C2$. Assuming that we wish to avoid the most disadvantageous cases, we will rank $C2$ higher than $C1$ so as to avoid the possibility which might by implied by $C1$, namely: $(0.6, 0.4, 0)$ (while the entire hesitation margin is added to the non-membership value). The best result which could happen (if the entire hesitation margin is added to the membership value of $C1$), namely $(0.8, 02, 0)$, (i.e. case $C6$ ranked as the best one – $R(C6) = 0.1$) does not influence the ranking of $C1$(3.232).

Analogous situation can be observed for the pairs: $C3$ and $C4$, and next for $C5$ and $C6$. It is easy to notice that the existence of the non-zero hesitation margin influences negatively the ranking.

The obtained results seem to meet our expectations pretty well.

Finally, we will verify the results produced by (3.232) with the normalized Euclidean distance (3.157).

At the beginning, we will rank the same alternatives using (3.232) as we have done previously using (3.231), i.e.: $Y_1=(0, 0.8, 0.2)$, $Y_2=(0, 0.6, 0.4)$, $Y_3=(0, 0.3, 0.7)$, and $Y_4=(0, 0, 1)$. We obtain $R_E(Y_1)=0.55$, $R_E(Y_2)=0.61$, $R_E(Y_3)=0.85$, $R_E(Y_4)=1$. The results seem to render our intuition now.

The results obtained via (3.232) for the most characteristic alternatives are still the same for the normalized Euclidean distance (3.157) as they were for the normalized Hamming distance (3.156). As previously (i.e., with the normalized Hamming distance (3.157)), the best is alternative $M(1,0,0)$ ($R_E(M) = 0$). For alternative $N(0,1,0)$, again, we obtain $R_E(N) = 0.5$ (N is fully reliable as the hesitation margin is equal 0 but the distance $e_{IFS}(M,N) = 1$). In general, on MN, the values of R_E decrease from 0.5 (for alternative N) to 0 (for the best alternative M). The maximal value of R_E, i.e. 1, is for $O(0,0,1)$ for which $e_{IFS}(M,O), \pi_O = 1$ (alternative O "indicates" the whole triangle MNO). All other alternatives Y_i "indicate" smaller triangles $Y_i Y_{i,min} Y_{i,max}$ (Figure 3.6), so that their R_E's are smaller (better as to the amount of reliable information).

It is worth emphasizing that the results obtained via (3.232), which reflect our intuition concerning ranking of the alternatives, are obtained using all three terms describing the intuitionistic fuzzy alternatives, i.e., membership values,

non-membership values, and the hesitation margin values. Also the distances (3.157)
in (3.232) are calculated taking into account all three terms. In other words, we use
a 3D representation of the intuitionistic fuzzy sets.

Moreover, the proposed measure (3.232) strongly emphasizes the difference between knowledge (represented by the membership and non-membership values) and
lack of knowledge (represented by the hesitation margins). Even if an alternative
does not fulfill our criteria at all (alternative N), it is ranked higher ($R_E(N) = 0.5$)
than an alternative about which we can say nothing (alternative O). Other examples
are given in Table 3.2 (Szmidt and Kacprzyk [208]).

Table 3.2 Examples of results showing that (3.232) reflects differences between negative
knowledge and lack of knowledge in the ranking of the alternatives

No.	Alternative (μ,ν,π)	$R_E(Y_i)$
1	$(0,0.8,0.2)$	0.550
2	$(0,0.2,0.8)$	0.825
3	$(0,0.7,0.3)$	0.578
4	$(0,0.3,0.7)$	0.755
5	$(0,0.6,0.4)$	0.610
6	$(0,0.4,0.6)$	0.697
7	$(0,1,0)$	0.5
8	$(0,0,1)$	1

The results provided in Table 3.2 make it possible to come to some conclusions
concerning the situations for which we have a fixed membership value of the alternatives (membership value is equal to 0 in Table 3.2), namely:

- an alternative is ranked lower (which means bigger values from (3.232)) the
  smaller the non-membership function and the bigger the hesitation margin (cf.
  the sequence of cases: 1, 3, 5, 8);
- an alternative is ranked higher (i.e., the smaller the values from (3.232)) the
  higher the non-membership function and the lower the hesitation margin (cf. the
  sequence of cases: 2, 4, 6, 7);
- "negative knowledge" represented by the non-membership values, and lack of
  knowledge represented by the hesitation margins are different from the point of
  view of (3.232) (cf. the pairs: 1 and 2, 3 and 4, 5 and 6, 7 and 8).

Other examples, presented in Table 3.3 (Szmidt and Kacprzyk [208]) make it
possible to notice that:

- an alternative is ranked higher (which means that the values from (3.232) are
  lower) for a fixed value of the non-membership function (cf. Table 3.3, the cases:
  2, 4, 6, 8, for which the non-membership value is equal 0) the higher the values
  of the membership function (lower hesitation margins);

Table 3.3 Examples of results showing that (3.232) reflects differences between positive knowledge and lack of knowledge in the ranking of the alternatives

No.	Alternative (μ, v, π)	$R_E(Y_i)$
1	$(0, 0.8, 0.2)$	0.550
2	$(0.8, 0, 0.2)$	0.12
3	$(0, 0.7, 0.3)$	0.578
4	$(0.7, 0, 0.3)$	0.195
5	$(0, 0.6, 0.4)$	0.610
6	$(0.6, 0, 0.4)$	0.280
7	$(0, 1, 0)$	0.5
8	$(1, 0, 0)$	0.

- the ranking function (3.232) does make a difference between the positive and negative knowledge (cf. Table 3.3, the pairs: 1 and 2, 3 and 4, 5 and 6, 7 and 8).

To sum up, the proposed ranking function (3.232) expresses differences both between knowledge and lack of knowledge, and between the positive and negative knowledge. In other words, the proposed function (3.232) seems to reflect the behavior of a human being in the process of ranking alternatives pretty well.

3.5   Concluding Remarks

We have considered distances between the intuitionistic fuzzy sets in two ways, employing:

– two term intuitionistic fuzzy set representation (membership values and non-membership values only were taken into account), and
– three term intuitionistic fuzzy set representation (membership values, non-membership values, and hesitation margins were taken into account).

We have discussed norms and metrics for both types of representations stressing their correctness from the mathematical point of view. However, the three term approach seems to be more justified and intuitively appealing from the practical point of view (which has its roots in some analytical and geometrical aspects).

Some problems have been shown concerning the Hausdorff distance while the Hamming metric was applied for the two term intuitionistic fuzzy set representation. It was shown, as well, that the method of calculating the Hausdorff distances in the same way which is correct for the interval-valued fuzzy sets does not work for the intuitionistic fuzzy sets.

Finally, the usefulness of the three term distances was emphasized in a measure of ranking of the intuitionistic fuzzy alternatives.

Chapter 4
Similarity Measures between Intuitionistic Fuzzy Sets

Abstract. In this chapter we consider similarity measures between intuitionistic fuzzy sets starting from reminding the axiomatic relation between distance and similarity measures. We show that this relation is not satisfied for the intuitionistic fuzzy sets. We also consider some similarity measures for the intuitionistic fuzzy sets, known from the literature. We show that neither similarity measures treating an intuitionistic fuzzy set as a simple interval-valued fuzzy set, nor straightforward generalizations of the similarity measures well-known for the classic fuzzy sets work under reasonable circumstances. Next, expanding upon our previous work, we consider a family of similarity measures constructed by taking into account both all the three terms (membership values, non-membership values, and hesitation margins) describing an intuitionistic fuzzy set, and the complements of the elements we compare. That is, we use all kinds and fine shades of information available. We also point out the traps one should be aware of while examining similarity between intuitionistic fuzzy sets. Finally, we consider correlation of the intuitionistic fuzzy sets.

4.1 Similarity Measures and Their Axiomatic Relation to Distance Measures

First we recall the well known relation of the distances and similarity measures for fuzzy sets (Liu [117]), and consider the correctness of the corresponding relation for the intuitionistic fuzzy sets.

Liu [117] proposed a definition of the axiomatic distance measure, and of the axiomatic similarity measure for fuzzy sets. The axioms make it possible to find the relationship between distance and similarity for fuzzy sets.

The distance measure d' between fuzzy sets (FS) proposed by Liu [117] is a real function $d' : FS \times FS \to R^+$ which for any fuzzy sets A', B', C' and a crisp set D has the following properties:

$$d'(A', B') = d'(B', A');$$
$$(4.1)$$

$$d'(A',A') = 0;   \tag{4.2}$$

$$d'(D,D^C) = \max_{A',B'} d'(A',B') \text{ if } D \text{ is a crisp set;}   \tag{4.3}$$

$$if A' \subset B' \subset C', \text{ then } d'(A',B') \le d'(A',C') \text{ and } d'(B',C') \le d'(A',C').   \tag{4.4}$$

The similarity measure s' between fuzzy sets (FS) proposed by Liu [117] is a real function $s' : FS \times FS \to R^+$ which for any fuzzy sets A', B', C' and a crisp set D has the following properties:

$$s'(A',B') = s'(B',A');   \tag{4.5}$$

$$s'(D,D^C) = 0 \text{ if } D \text{ is a crisp set;}   \tag{4.6}$$

$$s'(E,E) = \max_{A',B'} s'(A',B') \text{ if } E \text{ is a fuzzy set;}   \tag{4.7}$$

$$if A' \subset B' \subset C', \text{ then } s'(A',B') \ge s'(A',C') \text{ and } s'(B',C') \ge s'(A',C').   \tag{4.8}$$

After normalizing distance d', and similarity s', we have

$$0 \le d'(A',B') \le 1$$

and

$$0 \le s'(A',B') \le 1$$

for any fuzzy sets A' and B'. For normalized d' and s' we have (Liu [117]):

$$d' = 1 - s'.   \tag{4.9}$$

In other words, the distance d' and the similarity measure s' are dual concepts in the case of fuzzy sets.

The above approach, which gives correct results in the case of fuzzy sets, has been applied also for the intuitionistic fuzzy sets (cf. Hung and Yang [88]).

The distance measure d between intuitionistic fuzzy sets (IFS) proposed by Hung and Yang [88] is a real function $d : IFS \times IFS \to R^+$ which for any intuitionistic fuzzy sets A, B, C and a crisp set D has the following properties:

$$d(A,B) = d(B,A);   \tag{4.10}$$

$$d(A,A) = 0; \tag{4.11}$$

$$d(D,D^C) = \max_{A,B} d(A,B) \text{ if } D \text{ is a crisp set;} \tag{4.12}$$

$$if A \subset B \subset C, \text{ then } d(A,B) \le d(A,C) \text{ and } (B,C) \le d(A,C). \tag{4.13}$$

The similarity measure s between intuitionistic fuzzy sets (IFS) proposed by Hung and Yang [88] is a real function $s : IFS \times IFS \rightarrow R^+$ which for any intuitionistic fuzzy sets A and B and a crisp set D has the following properties:

$$s(A,B) = s(B,A); \tag{4.14}$$

$$s(D,D^C) = 0 \text{ if } D \text{ is a crisp set;} \tag{4.15}$$

$$s(E,E) = \max_{A,B} s(A,B) \text{ if } E \text{ is an intuitionistic fuzzy set;} \tag{4.16}$$

$$if A \subset B \subset C, \text{ then } s(A,B) \ge s(A,C) \text{ and } s(B,C) \ge s(A,C). \tag{4.17}$$

After normalizing distance d, and similarity s, we have

$$0 \le d(A,B) \le 1,$$

and

$$0 \le s(A,B) \le 1$$

for any intuitionistic fuzzy sets A and B, so it is easy to notice that

$$d = 1 - s. \tag{4.18}$$

In other words, the distance d and the similarity measure s are presented in literature (e.g., [88]) as dual concepts in the case of the intuitionistic fuzzy sets, too.

However, in Szmidt and Kacprzyk [205] it is shown that the (*1-Hamming distance*), where the normalized Hamming distance between the intuitionistic fuzzy sets is given by (3.156), should not be used as a similarity measure between them.

The similarity measure, corresponding to distance (3.156), is, due to (4.18), equal to:

$$Sim_H = 1 - l_{IFS}^1(A,B) = 1 - \frac{1}{2n}\sum_{i=1}^{n}(|\mu_A(x_i) - \mu_B(x_i)| +$$
$$+ |v_A(x_i) - v_B(x_i)| + |\pi_A(x_i) - \pi_B(x_i)|). \tag{4.19}$$

Figures 4.1 and 4.2 show that for a fixed membership value all the elements (with membership value equal to the fixed value) are at the same distance from the element $(1,0,0)$. The situation repeats when we examine distances to any element $x : (\mu, v, \pi)$ making use of the normalized Hamming distance. We obtain the same

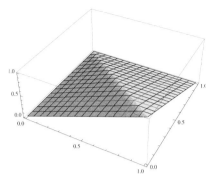

Fig. 4.1 Values of similarity (4.19) for any element from an intuitionistic fuzzy set and the element (1, 0, 0)

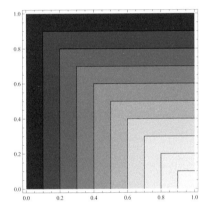

Fig. 4.2 Contourplot of (4.19) for any element from an intuitionistic fuzzy set and the element (1,0,0)

type of shapes (Figures 4.1 and 4.2) showing the elements equally distant from a fixed element.

Now we will demonstrate the result of using the (*1- normalized Euclidean distance*) as a similarity measure where the normalized Euclidean distance between the intuitionistic fuzzy sets is given by (3.157).

The similarity measure, corresponding to distance (3.157) is, due to (4.18), equal:

$$Sim_e(A,B) = 1 - q^1_{IFS}(A,B) = 1 - ((\frac{1}{2n}\sum_{i=1}^{n}(\mu_A(x_i) - \mu_B(x_i))^2 +$$

$$+ (\nu_A(x_i) - \nu_B(x_i))^2 + (\pi_A(x_i) - \pi_B(x_i))^2)^{\frac{1}{2}}. \qquad (4.20)$$

The results obtained from (4.20) are illustrated in Figures 4.3 and 4.4. Expressing similarity via distances means looking for geometric shapes, and while using (4.20), because of employing the Euclidean distance, we look in fact for the elements at a

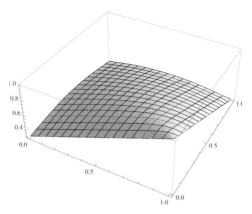

Fig. 4.3 Shape of $Sim_e(A,B)$ (4.20) for any element from an intuitionistic fuzzy set and $(1,0,0)$

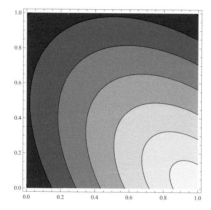

Fig. 4.4 Contourplot of (4.20) for any element from an intuitionistic fuzzy set and $(1,0,0)$

"radius" distance from a chosen element (object). It is obviously a correct approach (looking for some shapes in a coordinate space) but we should be rather careful while making conclusions about similarity as a dual measure of a distance as it is shown in the following example.

Example 4.1. For simplicity let us consider "degenerate", "one point type" in-tuitionistic fuzzy sets whose full description, including all three terms, is: $A = <x, \mu_A, \nu_A, \pi>/1$ exemplified by: M, N, L in $X = \{1\}$.

$$M = (1,0,0)/1, \quad N = (0,1,0)/1, \quad H = (0,0,1)/1$$

From (4.20) we obtain:
$Sim_e(M,N) = 0$, and
$Sim_e(M,H) = 0$
though N and H are obviously different. But the "the radius length" from M to N is

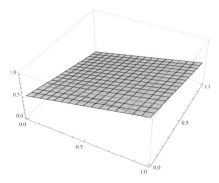

Fig. 4.5 Values obtained from (4.21) for any element from an intuitionistic fuzzy set and $(1, 0, 0)$

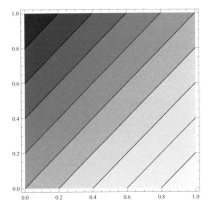

Fig. 4.6 Contourplot of (4.21) for any element from an intuitionistic fuzzy set and $(1, 0, 0)$

equal to the "radius length" from M to H. It is easy to accept for a crisp case that the elements on a circle are in the same distance from the middle of the circle which does not mean that all the elements belonging to the circle are "the same". Here we have the same situation.

We should also be cautious when considering similarity of the elements with a symmetry of terms in their description, e.g.:

$$M = (1,0,0)/1, \quad K = (0.5,0.3,0.2)/1,$$

$$L = (0.5,0.2,0.3)/1$$

for which the exchange between non-membership value and hesitation margin in K and L results in $Sim_e(M,K) = Sim_e(M,L)$ although for sure K and L are different but the "radiuses" MK and ML are the same.

We have shown previously in Chapter 3.3 (cf. also Szmidt and Kacprzyk [171], [188], [218]) that from the practical point of view it is necessary to take into account

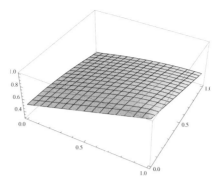

Fig. 4.7 Values obtained from (4.22) for any element from an intuitionistic fuzzy set and $(1, 0, 0)$

all three terms describing an intuitionistic fuzzy set while calculating distances. But it is interesting to verify the results for the "two term distances" between the intuitionistic fuzzy sets and their effects as the measures dual to similarity. We will examine the "two term" normalized Hamming and Euclidean distances.

The similarity measure corresponding to the normalized Hamming distance between the intuitionistic fuzzy sets A, B in $X = \{x_1, \ldots, x_n\}$ when we use two terms only in the set description is:

$$Sim_{H2D} = 1 - l_{IFS(2)}(A, B) = 1 - \frac{1}{2n}\sum_{i=1}^{n}(|\mu_A(x_i) - \mu_B(x_i)| + $$
$$+ |\nu_A(x_i) - \nu_B(x_i)|). \tag{4.21}$$

The similarity measure corresponding to the normalized Euclidean distance between the intuitionistic fuzzy sets A, B in $X = \{x_1, \ldots, x_n\}$ when we use two terms only in the set description is:

$$Sim_{e2D}(A, B) = 1 - q_{IFS(2)}(A, B) = 1 - (\frac{1}{2n}\sum_{i=1}^{n}(\mu_A(x_i) - \mu_B(x_i))^2 + $$
$$+ (\nu_A(x_i) - \nu_B(x_i))^2)^{\frac{1}{2}}. \tag{4.22}$$

By analyzing (4.22) it is easy again to give examples of different situations (from the point of view of decision making) for which we obtain from (4.22) the same values.

Example 4.2. Again, consider "degenerate", "one point type" intuitionistic fuzzy sets, whose full description is $A = <x, \mu_A, \nu_A, \pi>/1$ exemplified by: M, P, R in $X = \{1\}$.

$$M = (1, 0, 0)/1, \quad P = (0.5, 0.3, 0.2)/1, \quad R = (0.4, 0, 0.6)/1$$

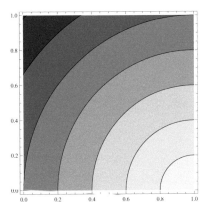

Fig. 4.8 Contourplot of (4.22) for any element from an intuitionistic fuzzy set and $(1,0,0)$

From (4.22) we obtain:
$Sim_{e2D}(M,P) = 0.7$, and
$Sim_{e2D}(M,R) = 0.7$
though P and R are obviously different. But the "the radius length" from M to P is equal to the "radius length" from M to R.

Figures 4.5–4.8 show that when we make use of the two term distances as dual concepts of similarity measures the situation does not change in the sense of the information obtained (certainly we do not suggest here that in general both ways of intuitionistic fuzzy set representations are equal, having in mind other drawbacks of the two term representation as compared to the three term representation of the intuitionistic fuzzy sets - cf. Chapter 3.3, Szmidt and Kacprzyk [218]).

Another similarity measure that is often used in practice is the cosine similarity measure which is based on Bhattacharya's distance [39], [148] and is expressed as the inner product of two vectors divided by the product of their lengths. In other words, it is the cosine of the angle between two vectors. The cosine similarity is often used in information retrieval [148]. Taking as a point of departure the three term intuitionistic fuzzy set representation, the cosine similarity measure is given by (4.23).

$$Sim_{mult}(A,B) = \frac{1}{n}\sum_{i=1}^{n}((\mu_A(x_i)\mu_B(x_i) + v_A(x_i)v_B(x_i) + \pi_A(x_i)\pi_B(x_i))/$$

$$/ (\mu_A(x_i)^2 + v_A(x_i)^2 + \pi_A(x_i)^2)^{\frac{1}{2}}(\mu_B(x_i)^2 + v_B(x_i)^2 + \pi_B(x_i)^2))^{\frac{1}{2}}) \qquad (4.23)$$

Figures 4.9 and 4.10 show values obtained from (4.23) illustrating similarity of the element $(1,0,0)$ and any other element belonging to an intuitionistic fuzzy set.

It is worth mentioning again that in (4.23) we might change the places of v and π, and the result of (4.23) will be the same although we consider quite different situations. This is clearly an undesired effect.

Example 4.3. Let us consider again "degenerate" intuitionistic fuzzy sets with full description (three term description) $A = \{< x, \mu_A, \nu_A, \pi > /1\}$, exemplified by: M, R, S in $X = \{1\}$, where
$M = (1, 0, 0)/1$, and

$$R = (0.5, 0.3, 0.2)/1, \quad S = (0.5, 0.2, 0.3)/1$$

From (4.23) we obtain:
$Sim_{mult}(M, R) = Sim_{mult}(M, S) = 0.81$, whereas R and S are obviously different, so that we assume that their similarity to the same M should be different, too.

Measure (4.23) cannot differentiate, e.g., between: $(0, 0, 1)$, and $(0, 1, 0)$ regarding their similarity to $(1, 0, 0)$ (cf. Figures 4.9 and 4.10). Certainly, we may point out many such cases with respect to (4.23).

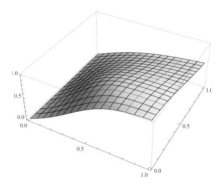

Fig. 4.9 Values obtained from measure (4.23) for any element from an intuitionistic fuzzy set and (1, 0, 0)

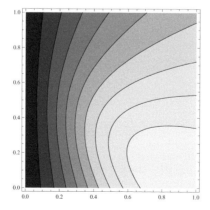

Fig. 4.10 Contourplot of measure (4.23) for any element from an intuitionistic fuzzy set and $(1, 0, 0)$

It is necessary to emphasize again that the above measures give correct answers in the sense of the formula used (geometrical shapes are recognized in respect to a chosen element) but in many situations we would expect that from the similarity measures we would be able at least to notice the existence of the complement element which seems to be the least similar to the considered element. This problem will be discussed in Section 4.3. Now we will verify other measures of similarity between the intuitionistic fuzzy sets, well known from the literature.

4.2 Some Other Counter-Intuitive Results Given by the Traditional Similarity Measures

There is a multitude of similarity measures both for the intuitionistic fuzzy sets (Atanassov [4, 6, 15]), and vague sets (Gau and Buehrer [72]) which have also been proved to be equivalent to the intuitionistic fuzzy sets (Bustine and Burillo [44]). Here we adopt the notation for the intuitionistic fuzzy sets but we will consider the measures originally introduced for vague sets, too. We consider the measures paying attention to the question whether the results produced are reliable.

One of the similarity measures between two intuitionistic fuzzy sets A and B was considered by Chen [51, 52], namely

$$S_C(A,B) = 1 - \frac{\sum_{i=1}^n |S_A(x_i) - S_B(x_i)|}{2n} \tag{4.24}$$

where: $S_A(x_i) = \mu_A(x_i) - \nu_A(x_i)$, $S_A(x_i) \in [-1,1]$ and $S_B(x_i) = \mu_B(x_i) - \nu_B(x_i)$, $S_B(x_i) \in [-1,1]$.

But, as Hong and Kim [83] noticed

$$\mu_A(x_i) - \nu_A(x_i) = \mu_B(x_i) - \nu_B(x_i) \Rightarrow S_C(A,B) = 1 \tag{4.25}$$

which is counterintuitive, since, e.g. [113], for $A = (x,0,0)$ and $B = (x,0.5,0.5)$, we have $S_C(A,B) = 1$.

Hong and Kim [83] proposed the similarity measures S_H and S_L to overcome the problem of S_C (4.24)

$$S_H(A,B) = 1 - (\sum_{i=1}^n |\mu_A(x_i) - \mu_B(x_i)| + |\nu_A(x_i) - \nu_B(x_i)|)/2n \tag{4.26}$$

$$S_L(A,B) = 1 - \frac{1}{4n}((\sum_{i=1}^n |S_A(x_i) - S_B(x_i)|) -$$

$$(\sum_{i=1}^n |\mu_A(x_i) - \mu_B(x_i)| + |\nu_A(x_i) - \nu_B(x_i)|)). \tag{4.27}$$

However, since $S_H(A,B)$ takes into account the absolute values, it does not distinguish between the positive and negative differences, e.g. (Li et al [113]):

for $A = \{(x, 0.3, 0.3)\}$, $B = \{(x, 0.4, 0.4)\}$, $C = \{(x, 0.3, 0.4)\}$, and $D = \{(x, 0.4, 0.3)\}$, we obtain from (4.26) that $S_H(A, B) = S_H(C, D) = 0.9$, which seems counter-intuitive.

$S_L(A, B)$ also gives counter-intuitive results, e.g. (Li et al [113]): for $A = \{(x, 0.4, 0.2)\}$, $B = \{(x, 0.5, 0.3)\}$, $C = \{(x, 0.5, 0.2)\}$, we obtain from (4.27) $S_L(A, B) = S_L(A, C) = 0.95$, which definitely seems counter-intuitive.

The same problem as with S_H occurs with the following similarity measure (cf. Li et al. [115]):

$$S_O(A, B) = 1 - (1/2n)^{0.5}(\sum_{i=1}^{n}(\mu_A(x_i) -$$

$$\mu_B(x_i))^2 + (\nu_A(x_i) - \nu_B(x_i))^2)^{0.5}. \tag{4.28}$$

Another similarity measure, considered by Dengfeng and Chuntian [111] was:

$$S_{DC}(A, B) = 1 - (1/n^{1/p})(\sum_{i=1}^{n}(|m_A(x_i) - m_B(x_i)|)^p)^{1/p} \tag{4.29}$$

where:
$m_A(x_i) = (\mu_A(x_i) + 1 - \nu_A(x_i))/2$, $m_B(x_i) = (\mu_B(x_i) + 1 - \nu_B(x_i))/2$, $1 \leq p < \infty$. Unfortunately, since in (4.29) the medians of two intervals are compared only, it is rather easy to point out the counter-intuitive examples, e.g., $A = (x, 0.4, 0.2)$, $B = (x, 0.5, 0.3)$, so that $S_{DC}(A, B) = 1$, for each p.

Dengfeng and Chuntian's measure S_{DC} (4.29) was modified by Mitchell [123], who applied a statistical approach by interpreting the intuitionistic fuzzy sets as families of ordered fuzzy sets. Let $\rho_\mu(A, B)$ and $\rho_\nu(A, B)$ denote a similarity measure between the "low" membership functions μ_A and μ_B, and between the "high" membership functions $1 - \nu_A$ and $1 - \nu_B$, respectively, as:

$$\rho_\mu(A, B) = S_{DC}(\mu_A, \mu_B) = 1 - (1/n^{1/p})(\sum_{i=1}^{n}(|\mu_A(x_i) - \mu_B(x_i)|)^p)^{1/p}$$

$$\rho_\nu(A, B) = S_{DC}(1 - \nu_A, 1 - \nu_B) = 1 - (1/n^{1/p})(\sum_{i=1}^{n}(|\nu_A(x_i) - \nu_B(x_i)|)^p)^{1/p},$$

then the modified similarity measure between A and B is

$$S_{HB}(A, B) = (\rho_\mu(A, B) + \rho_\nu(A, B))/2. \tag{4.30}$$

Unfortunately, S_{HB} gives the same counter-intuitive results as S_H, for $p = 1$ and for one-element sets.

Liang and Shi [116] proposed other measures, namely, $S_e^p(A, B)$, $S_s^p(A, B)$, and $S_h^p(A, B)$ to overcome the drawbacks of S_{DC}:

$$S_e^p(A, B) = 1 - (1/n^{1/p})(\sum_{i=1}^{n}(\phi_{\mu AB}(x_i) - \phi_{\nu AB}(x_i)|)^p)^{1/p} \tag{4.31}$$

where: $\phi_{\mu AB}(x_i) = |\mu_A(x_i) - \mu_B(x_i)|/2$,
$\phi_{\nu AB}(x_i) = |(1 - \nu_A(x_i))/2 - (1 - \nu_B(x_i))/2|$.
Unfortunately, for $p = 1$ and for one-element sets, $S_e^p(A,B) = S_{HB} = S_H$, which are
again the same counter-intuitive results.

$$S_s^p(A,B) = 1 - (1/n^{1/p})(\sum_{i=1}^{n}(\varphi_{s1}(x_i) - \varphi_{s2}(x_i))^p)^{1/p} \qquad (4.32)$$

where: $\varphi_{s1}(x_i) = |m_{A1}(x_t) - m_{B1}(x_i)|/2$,
$\varphi_{s2}(x_i) = |m_{A2}(x_i) - m_{B2}(x_i)|/2$,
$m_{A1}(x_i) = (\mu_A(x_i) + m_A(x_i))/2$,
$m_{A2}(x_i) = (m_A(x_i) + 1 - \nu_A(x_i))/2$,
$m_{B1}(x_i) = (\mu_B(x_i) + m_B(x_i))/2$,
$m_{B2}(x_i) = (m_B(x_i) + 1 - \nu_B(x_i))/2$,
$m_A(x_i) = (\mu_A(x_i) + 1 - \nu_A(x_i))/2$,
$m_B(x_i) = (\mu_B(x_i) + 1 - \nu_B(x_i))/2$.

Measure S_s^p (4.32) does not produce the problematic results like those obtained from S_{DC} (4.29) (for the intervals with equal medians) but, again, a problem of counter-intuitive results remains. For example (Li et al. [113]), for $A =
\{(x, 0.4, 0.2)\}, B = \{(x, 0.5, 0.3)\}, C = \{(x, 0.5, 0.2)\}$, we obtain $S_s^p(A,B) = S_s^p(A,C)
= 0.95$ which seems difficult to accept.
    S_h^p is given as

$$S_h^p(A,B) = 1 - (1/(3n)^{1/p})(\sum_{i=1}^{n}(\eta_1(i) + \eta_2(i) + \eta_3(i))^p)^{1/p} \qquad (4.33)$$

where:

$\eta_1(i) = \phi_\mu(x_i) + \phi_\nu(x_i)$ (the same as for S_e^p),
$\eta_2(i) = m_A(x_i) - m_B(x_i))$ (the same as for S_{DC}),
$\eta_3(i) = \max(l_A(i), l_B(i)) - \min(l_A(i), l_B(i))$,
$l_A(i) = (1 - \nu_A(x_i) - \mu_A(x_i))/2$,
$l_B(i) = (1 - \nu_B(x_i) - \mu_B(x_i))/2$.

However, again, this measure also does not avoid the counter-intuitive cases. For
$A = (x, 0.3, 0.4)$, and $B = (x, 0.4, 0.3)$, i.e., for an element and its complement, we
obtain $S_h^p(A,B) = 0.933$ (which seems to be rather too big a similarity for an element
and its complement).
    The similarity measures S_{HY}^1, S_{HY}^2, S_{HY}^3, in which Hausdorff distances are em-
ployed, were proposed by Hung and Yang [87]:

$$S_{HY}^1(A,B) = 1 - d_H(A,B) \qquad (4.34)$$

$$S_{HY}^2(A,B) = 1 - (e^{d_H(A,B)} - e^{-1})/(1 - e^{-1}) \qquad (4.35)$$

$$S_{HY}^3(A,B) = (1 - d_H(A,B))/(1 + d_H(A,B)) \qquad (4.36)$$

where:

$d_H(A,B)) = \sum_{i=1}^n \max(|\mu_A(x_i) - \mu_B(x_i)|, |v_A(x_i) - v_B(x_i)|)$.

But (4.34)–(4.36) do also give counter-intuitive results (implied by the calculation of the distance $d_H(A,B)$ – cf. Section 3.3.2.2, and Szmidt and Kacprzyk [218]). For example, if $A = (x, 0.4, 0.5)$, $B = (x, 0.5, 0.4)$, $C = (x, 0.5, 0.3)$, $D = (x, 0.6, 0.4)$, $E = (x, 0.6, 0.3)$, $F = (x, 0.4, 0.3)$ then $S_{HY}^1(A,B) = 0.9$ (a counter-intuitive large similarity for A and its complement as $B = A^C$), and also $S_{HY}^1(C,D) = S_{HY}^1(C,E) = S_{HY}^1(C,F) = 0.9$. Next, $S_{HY}^2(A,B) = S_{HY}^2(C,D) = S_{HY}^2(C,E) = S_{HY}^2(C,F) = 0.85$, and also $S_{HY}^3(A,B) = S_{HY}^3(C,D) = S_{HY}^3(C,E) = S_{HY}^3(C,F) = 0.85$.

Hung and Yang [88] made a straightforward attempt to calculate the similarity between the intuitionistic fuzzy sets just by adding the non-memberships values to the existing similarity measures for fuzzy sets. Their measures (4.37) and (4.38) are the extension of Wang's measures [241]:

$$S_{w1}(A,B) = (1/n) \sum_{i=1}^n \frac{\min(\mu_A(x_i), \mu_B(x_i)) + \min(v_A(x_i), v_B(x_i))}{\max(\mu_A(x_i), \mu_B(x_i)) + \max(v_A(x_i), v_B(x_i))}. \qquad (4.37)$$

But again, it is easy to give counter-examples. For example, for $A = \{(x, 0, 0.5)\}$, $B = \{(x, 0.1, 0.5)\}$, $C = \{(x, 0, 0.6)\}$, we obtain from (4.37): $S_{w1}(A,B) = S_{w1}(A,C) = 0.8(3)$ (for different B and C we obtain the same result), for $A = \{(x, 0, 0.5)\}$, $B = \{(x, 0.18, 0.5)\}$, $C = \{(x, 0, 0.68)\}$, we obtain $S_{w1}(A,B) = S_{w1}(A,C) = 0.735$ (again - for different B and C the same result) etc., which seems to be difficult to accept (S_{w1} is not bijective).

$$S_{w2}(A,B) = (1/n) \sum_{i=1}^n (1 - 0.5(|\mu_A(x_i) - \mu_B(x_i)| + |v_A(x_i) - v_B(x_i)|)) \qquad (4.38)$$

However, for $A = \{(x, 0, 0.5)\}$, $B = \{(x, 0, 0.4)\}$, $C = \{(x, 0, 0.6)\}$, we obtain: $S_{w2}(A,B) = S_{w2}(A,C) = 0.95$ (again - for different B and C the same similarity), which seems to be difficult to accept (S_{w2} is not bijective).

Three extensions of Pappis and Karacapilidis' [129] similarity measures for fuzzy sets were proposed by Hung and Yang [88]. The introduced measures (4.39), (4.40) and (4.41), proposed for the intuitionistic fuzzy sets, are straightforward extensions of fuzzy similarity measures.

$$S_{pk1}(A,B) = \frac{\sum_{i=1}^n (\min(\mu_A(x_i), \mu_B(x_i)) + \min(v_A(x_i), v_B(x_i)))}{\sum_{i=1}^n (\max(\mu_A(x_i), \mu_B(x_i)) + \max(v_A(x_i), v_B(x_i)))}, \qquad (4.39)$$

for which the counter-intuitive examples are the same as for (4.37).

$$S_{pk2}(A,B) = 1 - 0.5(\max_i(|\mu_A(x_i) - \mu_B(x_i)|) +$$

$$\max_i(|v_A(x_i) - v_B(x_i)|)) \qquad (4.40)$$

It is easy to give counter-examples for (4.40), which is especially well visible for one-element sets. For example, for $A = \{(x,0,0.5)\}$, $B = \{(x,0.1,0.5)\}$, $C = \{(x,0,0.6)\}$, we obtain $S_{pk2}(A,B) = S_{pk2}(A,C) = 0.95$ (for different B and C just the same result).

$$S_{pk3}(A,B) = 1 - \frac{\sum_{i=1}^{n}(|\mu_A(x_i) - \mu_B(x_i)| + |\nu_A(x_i) - \nu_B(x_i)|)}{\sum_{i=1}^{n}(|\mu_A(x_i) + \mu_B(x_i)| + |\nu_A(x_i) + \nu_B(x_i)|)}, \qquad (4.41)$$

but again, for $A = \{(x,0,0.5)\}$, $B = \{(x,0,0.26)\}$, $C = \{(x,0,0.965)\}$, we obtain $S_{pks}(A,B) = S_{pk3}(A,C) = 0.68$ (for different B and C we obtain the same similarity), which seems to be difficult to accept.

4.2.1 Why the Measures Presented May Yield Counter-Intuitive Results?

All the similarity measures presented above at first glance seem to be different. However, all o them were constructed to satisfy the following conditions:

$$S(A,B) \in [0,1] \qquad (4.42)$$
$$S(A,B) = 1 \Longleftrightarrow A = B \qquad (4.43)$$
$$S(A,B) = S(B,A) \qquad (4.44)$$
$$If\ A \subseteq B \subseteq C,\ then\ S(A,C) \leq S(A,B)\ and$$
$$S(A,C) \leq S(B,C) \qquad (4.45)$$

Conditions (4.42)–(4.44) are obvious. The problem lies in (4.45) as this condition is meant as:

$$A \subset B\ iff\ \forall x \in X, \quad \mu_A(x) \leq \mu_B(x)\ and$$
$$\nu_A(x) \geq \nu_B(x) \qquad (4.46)$$

It is worth emphasizing that condition (4.46) is not constructive and operational for the intuitionistic fuzzy sets as for many cases it can not be used. For example, for the elements $(x : (\mu, \nu, \pi))$:
x_1: $(0.12,0.4,0.48)$ and x_2: $(0.1,0.3,0.6)$ we can not come to a conclusion. Moreover, element x: $(0,0,1)$ seems to be always beyond consideration in the sense of (4.46) which is very specific, and mostly practically irrelevant.

Next, the measures presented above were constructed making use of two term representation of the intuitionistic fuzzy sets – only the membership values and non-membership values were taken into account. It is worth noticing that using the two term representation is equivalent to representing an intuitionistic fuzzy set by an interval (in several discussed above similarity measures just two intervals were compared – each interval representing one of the intuitionistic fuzzy sets under comparison). But if we bear in mind that the elements of an intuitionistic fuzzy set are described via the membership values, non-membership values, and the hesitation margins, i.e., accepting the three term description, comparison of two intervals only is not enough. Making use of the three term description of the intuitionistic fuzzy sets, in the terms of intervals, we have both the membership value in an interval, and the non-membership value in an interval so that we should represent an intuitionistic fuzzy set via two (not one) intervals.

Considering the two term representation of the intuitionistic fuzzy sets, which is equivalent to representing the intuitionistic fuzzy sets as single intervals implies some problems while calculating distances. Distances used in the (counter-intuitive) similarity measures mentioned in the previous section are calculated without taking into account the hesitation margins as the membership and non-membership values only are taken into account. The counter-intuitive results obtained in such situations are discussed in Chapter 3, as well as in Szmidt and Kacprzyk [171], [188], [218].

4.3   An Example of Intuitively Justified and Operational Similarity Measure

First we recall the measure of similarity between the intuitionistic fuzzy sets presented by Szmidt and Kacprzyk [181], [180]).

Let us calculate the similarity of any two elements belonging to an intuitionistic fuzzy set (sets) which are geometrically represented by points X and F (Figure 4.11) belonging to the triangle MNH. The proposed measures indicate whether X is more similar to F or to F^C, where F^C is the complement of F. In other words, the proposed measures answer the question if X is more similar or more dissimilar to F (Figure 4.11), expressed as:

$$Sim_{rule}(X,F) = \frac{l^1_{IFS}(X,F)}{l^1_{IFS}(X,F^C)} \qquad (4.47)$$

where: $l^1_{IFS}(X,F)$ is a distance from $X(\mu_X, \nu_X, \pi_X)$ to $F(\mu_F, \nu_F, \pi_F)$,
$l^1_{IFS}(X,F^C)$ is a distance from $X(\mu_X, \nu_X, \pi_X)$ to $F^C(\nu_F, \mu_F, \pi_F)$,
F^C is a complement of F, distances $l^1_{IFS}(X,F)$ and $l^1_{IFS}(X,F^C)$ are calculated from (3.156).

The following conditions are fulfilled for (4.47)

$$0 \leq Sim_{rule}(X,F) \leq \infty \qquad (4.48)$$

$$Sim_{rule}(X,F) = Sim_{rule}(F,X) \qquad (4.49)$$

The similarity has typically been assumed to be symmetric. Tversky [237], however, has provided some empirical evidence that similarity should not always be treated as a symmetric relation. We recall this to show that a similarity measure (4.47) may have some features which can be useful in some situations but are not welcome in others (see Cross and Sudkamp [55], Wang et al. [244], Veltkamp [239]).

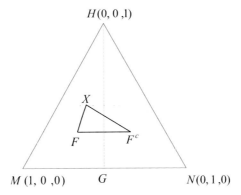

Fig. 4.11 The triangle *MNH* explaining the ratio-based measure of similarity

Szmidt and Kacprzyk [180] have noticed that the formula (4.47) can also be
stated as

$$Sim_{rule}(X,F) = \frac{l_{IFS}(X,F)}{l_{IFS}(X,F^C)} = \frac{l_{IFS}(X^C,F^C)}{l_{IFS}(X,F^C)} =$$

$$= \frac{l_{IFS}(X,F)}{l_{IFS}(X^C,F)} = \frac{l_{IFS}(X^C,F^C)}{l_{IFS}(X^C,F)} \qquad (4.50)$$

It can be noticed that

- $Sim_{rule}(X,F) = 0$ means that X and F are identical.
- $Sim_{rule}(X,F) = 1$ means that X is to the same extent similar to F and F^C, whereas
  the values bigger than 1 mean a closer similarity of X and F^C to X and F.
- When $X = F^C$ (or $X^C = F$), there is $l_{IFS}^1(X,F^C) = l_{IFS}^1(X^C,F) = 0$ meaning the
  complete dissimilarity of X and F (or in other words, the identity of X and F^C),
  and then $Sim_{rule}(X,F) \to \infty$.
- $X = F = F^C$ means the highest possible entropy (see [175]) for both elements F
  and X i.e. the highest "fuzziness" – not too constructive a case when looking for
  compatibility (both similarity and dissimilarity).

The above properties mean that while applying the measure (4.47) to analyze the
similarity of two objects, one should be interested in the values
$0 \leq Sim_{rule}(X,F) < 1$.

The proposed measure (4.47) was constructed for selecting objects which are
more similar than dissimilar [and well-defined in the sense of possessing (or not)
attributes we are interested in].

Now we will show that a measure of similarity defined as above, (4.47), between
$X(\mu_X, \nu_X, \pi_X)$ and $F(\mu_F, \nu_F, \pi_F)$ is more powerful than a simple distance between
them.

Example 4.4. (Szmidt and Kacprzyk [181])
 Let X and F be two elements belonging to an intuitionistic fuzzy set (with the coordinates (μ, v, π)),

$$X = (0.5, 0.4, 0.1)$$

$$F = (0.4, 0.5, 0.1)$$

so that

$$F^C = (0.5, 0.4, 0.1)$$

and from (4.47) we have

$$l^1_{IFS}(X, F) = \frac{1}{2}(|0.5 - 0.4| + |0.4 - 0.5| + |0.1 - 0.1|) = 0.1 \qquad (4.51)$$

which means that the distance is small – and just taking this into account, we would say that X and F are similar. However

$$l^1_{IFS}(X, F^C) = \frac{1}{2}(|0.5 - 0.5| + |0.4 - 0.4| + |0.1 - 0.1|) = 0 \qquad (4.52)$$

which means that X is just the same as F^C. We can not speak at all about similarity of X and F despite the fact that the distance between them is small.   □

   It is worth stressing again that a measure of similarity defined as above, (4.47), between $X(\mu_X, v_X, \pi_X)$ and $F(\mu_F, v_F, \pi_F)$ is more powerful than a simple distance between them.
   So to sum up:

- a big distance between two (or more) objects/elements, or sets means for sure that the similarity does not occur.
- when a distance is small, we can say nothing for sure about similarity just on the basis of a distance between two objects [when not taking into account complements of the objects as in (4.47)]. The distance between objects can be small and the compared objects can be more dissimilar than similar.

4.3.1   Analysis of Agreement in a Group of Experts

Now, we will use the similarity measure (4.47) to evaluate the extent of agreement between experts. The concept of consensus, as stressed by Loewer and Laddaga [120], i.e. a full and unanimous agreement, should be softened because in practice consensus means that "most of the individuals agree as to most of the options". Many works have been published along this line, notably in Kacprzyk and Fedrizzi [93, 94] a new measure of consensus was proposed. Here we follow that line of reasoning but in the intuitionistic fuzzy setting, using a different perspective based on the use of a similarity measure.
   So, if all of the considered pairs of experts' preferences are

- just the same (i.e. full agreement meaning consensus in a traditional sense) – the proposed measure of similarity (4.47) is equal to 0,
- quite opposite (i.e. full disagreement) – similarity (4.47) tends to infinity,
- different to some extent – distance from the consensus belongs to the open interval $(0,1)$
- to the same extent similar as dissimilar – the proposed measure (4.47) is equal to 1.

Preferences given by each individual are expressed via intuitionistic fuzzy sets (describing intuitionistic fuzzy preferences). Having in mind that distances between intuitionistic fuzzy sets should be calculated taking into account all three terms characterizing an intuitionistic fuzzy set, we start from a set of data which consists of three types of matrices describing individual preferences. The first type of matrices is the same as for classical fuzzy sets, i.e. membership functions $[r_{ij}^k]$ given by each individual k concerning each pair of options ij. But, additionally, it is necessary to take into account hesitation margins $[\pi_{ij}^k]$ and non-membership functions $[v_{ij}^k]$.

Generally, the extent of similarity for two experts k_1, k_2 considering n options is given as (Szmidt and Kacprzyk [181])

$$Sim^{k_1,k_2} = \frac{1}{A} \sum_{i=1}^{n-1} \sum_{j=i+1}^{n} Sim^{k_1,k_2}(i,j) =$$

$$= \frac{1}{A} [\sum_{i=1}^{n-1} \sum_{j=i+1}^{n} (\mid \mu_{ij}(k_1) - \mu_{ij}(k_2) \mid +$$

$$+ \mid v_{ij}(k_1) - v_{ij}(k_2) \mid + \mid \pi_{ij}(k_1) - \pi_{ij}(k_2) \mid)] /$$

$$/ [\sum_{i=1}^{n-1} \sum_{j=i+1}^{n} (\mid \mu_{ij}(k_1) - v_{ij}(k_2) \mid +$$

$$+ \mid v_{ij}(k_1) - \mu_{ij}(k_2) \mid + \mid \pi_{ij}(k_1) - \pi_{ij}(k_2) \mid)] \tag{4.53}$$

where

$$A = \frac{1}{2C_n^2} = \frac{1}{n(n-1)} \tag{4.54}$$

For m experts, we examine similarity of their preferences in pairs (4.53) and next, we find an agreement of all experts

$$Sim = \frac{1}{m(m-1)} \sum_{p=1}^{m-1} \sum_{r=p+1}^{m} Sim^{k_p,k_r} \tag{4.55}$$

where Sim^{k_p,k_r} is given by (4.53).

Example 4.5. (Szmidt and Kacprzyk [181]) Suppose that there are 3 individuals ($m = 3$) considering 3 options ($n = 3$), and the individual intuitionistic fuzzy preference relations are:

$$\mu^1(i,j) = \begin{bmatrix} - & .1 & .5 \\ .9 & - & .5 \\ .4 & .3 & - \end{bmatrix} \quad v^1(i,j) = \begin{bmatrix} - & .9 & .4 \\ .1 & - & .3 \\ .5 & .5 & - \end{bmatrix} \quad \pi^1(i,j) = \begin{bmatrix} - & 0 & .1 \\ 0 & - & .2 \\ .1 & .2 & - \end{bmatrix}$$

$$\mu^2(i,j) = \begin{bmatrix} - & .1 & .5 \\ .9 & - & .5 \\ .2 & .2 & - \end{bmatrix} \quad v^2(i,j) = \begin{bmatrix} - & .9 & .2 \\ .1 & - & .2 \\ .5 & .5 & - \end{bmatrix} \quad \pi^2(i,j) = \begin{bmatrix} - & 0 & .3 \\ 0 & - & .3 \\ .3 & .3 & - \end{bmatrix}$$

$$\mu^3(i,j) = \begin{bmatrix} - & .2 & .1 \\ .8 & - & .6 \\ .2 & .3 & - \end{bmatrix} \quad v^3(i,j) = \begin{bmatrix} - & .8 & .2 \\ .2 & - & .3 \\ .1 & .6 & - \end{bmatrix} \quad \pi^3(i,j) = \begin{bmatrix} - & 0 & .7 \\ 0 & - & .1 \\ .7 & .1 & - \end{bmatrix}$$

It is worth noticing that the following conditions for each pair of options (i,j) and expert k should be fulfilled:

$$\mu^k(i,j) = v^k(j,i) \tag{4.56}$$

which guarantee that
− $r^k(i,j) + r^k(j,i) = 1$ is obtained in the particular fuzzy case, and
− the natural relation $l^{p,r}(i,j) = l^{p,r}(j,i)$ for each pair of experts is satisfied in this way.

To find the extent of agreement in the group, we must calculate similarity $Sim^{p,r}(i,j)$ for each pair of experts (p,r) considering each pair of options (i,j).

First, we calculate similarity for each pair of experts concerning the first and the second option. For example, the data and the calculations for the second and the third experts are (Szmidt and Kacprzyk [181]):
$F^2(1,2) = (0.1,0.9,0)$ - preferences of the second expert,
$F^3(1,2) = (0.2,0.8,0)$ - preferences of the third expert,
$F^{3,C}(1,2) = (0.8,0.2,0)$ - the complement of $F^3(1,2)$, i.e., opposite preferences of the third expert.

From (4.47) and (4.53) we have

$$Sim^{2,3}(1,2) = \frac{l(F^2(1,2),F^3(1,2))}{l(F^2(1,2),F^{3,C}(1,2))} = 0.14 \tag{4.57}$$

For experts $(1,2)$ and $(1,3)$ we obtain

$$Sim^{1,2}(1,2) = \frac{l(F^1(1,2),F^2(1,2))}{l(F^1(1,2),F^{2,C}(1,2))} = 0 \tag{4.58}$$

$$Sim^{1,3}(1,2) = \frac{l(F^1(1,2),F^3(1,2))}{l(F^1(1,2),F^{3,C}(1,2))} = 0.2 \tag{4.59}$$

The average similarities for the three considered experts considering options $(1,2)$ are obtained from (4.57)-(4.59), namely

$$Sim(1,2) = \frac{1}{3}(0 + 0.2 + 0.14) = 0.11 \tag{4.60}$$

For options $(1,3)$ we obtain the following results

$$Sim^{1,2}(1,3) = \frac{l(F^1(1,3), F^2(1,3))}{l(F^1(1,3), F^{2,C}(1,3))} = 0.67 \tag{4.61}$$

$$Sim^{1,3}(1,3) = \frac{l(F^1(1,3), F^3(1,3))}{l(F^1(1,3), F^{3,C}(1,3))} = 1 \tag{4.62}$$

$$Sim^{2,3}(1,3) = \frac{l(F^2(1,3), F^3(1,3))}{l(F^2(1,3), F^{3,C}(1,3))} = 1 \tag{4.63}$$

Aggregating the above values we obtain the similarity for options $(1,3)$

$$Sim(1,3) = \frac{1}{3}(0.67 + 1 + 1) = 0.89 \tag{4.64}$$

Finally, for options $(2,3)$ we have

$$Sim^{1,2}(2,3) = \frac{l(F^1(2,3), F^2(2,3))}{l(F^1(2,3), F^{2,C}(2,3))} = 0.33 \tag{4.65}$$

$$Sim^{1,3}(2,3) = \frac{l(F^1(2,3), F^3(2,3))}{l(F^1(2,3), F^{3,C}(2,3))} = 0.33 \tag{4.66}$$

$$Sim^{2,3}(2,3) = \frac{l(F^2(2,3), F^3(2,3))}{l(F^2(2,3), F^{3,C}(2,3))} = 0.57 \tag{4.67}$$

Aggregating the above values we obtain the similarity for options $(2,3)$

$$Sim(2,3) = \frac{1}{3}(0.33 + 0.33 + 0.57) = 0.41 \tag{4.68}$$

The above results show that the biggest agreement in our group concerns options $(1,2)$ – the similarity measure is equal to 0.11. The smallest agreement concerns options $(1,3)$ for which the similarity measure is equal to 0.89.

Certainly, similar calculations can be performed for experts (aggregation is performed by experts). The results are (Szmidt and Kacprzyk [181]):

$$Sim^{1,2} = 0.33 \tag{4.69}$$

$$Sim^{1,3} = 0.51 \tag{4.70}$$

$$Sim^{2,3} = 0.57 \qquad (4.71)$$

The most similar are the preferences of the first and the second expert (4.69), the least similar are the preferences of the second and the third experts (4.71).

The similarity measure aggregated both by options and by experts, i.e., the general similarity for the group, is obtained by the aggregation of results (4.69)-(4.71)

$$Sim = \frac{1}{3}(0.33 + 0.51 + 0.57) = 0.47 \qquad (4.72)$$

Certainly, just the same results will be obtained when aggregating (4.60), (4.64) and (4.68).

In our example the agreement of the group of experts (the similarity concerning all options) is equal to 0.47 (not bad). □

In the method of group agreement analysis, presented above, it is possible to take into account the fact that some experts can be more important than others – proper weights for pairs of individuals can be taken into account in formula (4.55).

4.4 More Examples of Similarity Measures Including the Notion of Complement

In the previous section we have shown on a simple example (see also Szmidt and Kacprzyk [181]) that the measure (4.47) gives reasonable results when applied to assessing agreement in a group of experts. However, the measure has its disadvantage, namely, it does not follow the range of the usually assumed values for the similarity measures. It is possible, though, to construct a whole array of similarity measures following the philosophy, and preserving the advantages of the measure (4.47), and whose numerical values are consistent with the common scientific tradition (i.e. belonging to $[0,1]$).

To be more specific, when constructing the new similarity measures we used the same two kinds of distances as in (4.47) (i.e., $l^1_{IFS}(X,F)$, $l^1_{IFS}(X,F^C)$) but we were looking for a function with values from $[0,1]$. Specifically, following (Szmidt and Kacprzyk [193])

$$f(l^1_{IFS}(X,F), l^1_{IFS}(X,F^C)) = \frac{l^1_{IFS}(X,F)}{l^1_{IFS}(X,F) + l^1_{IFS}(X,F^C)} \qquad (4.73)$$

In (Szmidt and Kacprzyk [193]) it is stressed that the above function is constructed under the condition that case when $X = F = F^C$ (which is, by obvious reasons, not interesting in practice) was excluded from the considerations. The assumption $X = F = F^C$ means that one tries to compare an element (represented by) X about which nothing is known, to another element about which nothing is known. $F = F^C$ in terms of geometrical representation in Figure 4.11 means that X, F and F^C representing respective elements from an intuitionistic fuzzy set are at the same point on the HG segment. So the cases for which $l^1_{IFS}(X,F) = l^1_{IFS}(X,F^C) = 0$ were excluded from the considerations. Other cases are presented in Table 4.1.

Table 4.1 Possible values of (4.73) $c,d \in (0,1)$

$l^1_{IFS}(X,F)$		$l^1_{IFS}(X,F^C)$	f
0		1	0
1		0	1
1		1	0.5
c	less than	d	c/(c+d)<0.5
c	bigger than	d	d/(c+d)>0.5
c	equal to	d	0.5

In this way a function has been constructed which takes into account the same two distances as the previous measure (4.47) but now the new measure is normalized (its values are in $[0,1]$) (Szmidt and Kacprzyk [193]). It is obvious (see Table 4.1) that (4.73) is a concept dual to a similarity measure (if (4.73) is equal to zero then similarity is equal to 1; if (4.73) is equal to 1 then similarity is equal to zero). In other words, we may use (4.73) to construct a similarity measure.

As

$$0 \leq f(l^1_{IFS}(X,F), l^1_{IFS}(X,F^C)) \leq 1 \tag{4.74}$$

we would like to find a monotone decreasing function g fulfilling:

$$g(1) \leq g(f(l^1_{IFS}(X,F), l^1_{IFS}(X,F^C))) \leq g(0) \tag{4.75}$$

which means that

$$0 \leq g(f(l^1_{IFS}(X,F), l^1_{IFS}(X,F^C))) - g(1) \leq g(0) - g(1) \tag{4.76}$$

$$0 \leq \frac{g(f(l^1_{IFS}(X,F), l^1_{IFS}(X,F^C))) - g(1)}{g(0) - g(1)} \leq 1 \tag{4.77}$$

In this way we obtain a function having the properties of a similarity measure in a sense that it is monotone decreasing function of (4.73).

Definition 4.1. (Szmidt and Kacprzyk [193])

$$Sim(l^1_{IFS}(X,F), l^1_{IFS}(X,F^C)) = \frac{g(f(l^1_{IFS}(X,F), l^1_{IFS}(X,F^C))) - g(1)}{g(0) - g(1)} \tag{4.78}$$

where $f(l^1_{IFS}(X,F), l^1_{IFS}(X,F^C))$ is given by (4.73)

The simplest function g which may be applied in both definitions is

$$g(x) = 1 - x \tag{4.79}$$

which gives, from (4.78) (Szmidt and Kacprzyk [193]),

$$Sim_1(X,F) = Sim(l^1_{IFS}(X,F), l^1_{IFS}(X,F^C)) =$$

$$= 1 - f(l^1_{IFS}(X,F), l^1_{IFS}(X,F^C)) = 1 - \frac{l^1_{IFS}(X,F)}{l^1_{IFS}(X,F) + l^1_{IFS}(X,F^C)} \quad (4.80)$$

Another function $g(x)$ could be

$$g(x) = \frac{1}{1+x} \quad (4.81)$$

giving (Szmidt and Kacprzyk [193])

$$Sim_2(X,F) = Sim(l^1_{IFS}(X,F), l^1_{IFS}(X,F^C)) =$$

$$= \frac{1 - f(l^1_{IFS}(X,F), l^1_{IFS}(X,F^C))}{1 + f(l^1_{IFS}(X,F), l^1_{IFS}(X,F^C))} \quad (4.82)$$

Function

$$g(x) = \frac{1}{1+x^2} \quad (4.83)$$

gives (Szmidt and Kacprzyk [193])

$$Sim_3(X,F) = Sim(l^1_{IFS}(X,F), l^1_{IFS}(X,F^C)) =$$

$$= \frac{1 - f(l^1_{IFS}(X,F), l^1_{IFS}(X,F^C))^2}{1 + f(l^1_{IFS}(X,F), l^1_{IFS}(X,F^C))^2} \quad (4.84)$$

Theoretically, we could use as well $g(x) = \frac{1}{1+x^n}$ where $n = 3,4,\dots,k$ but the counterpart similarity measures $(\frac{1-x^n}{1+x^n})$ give the values which are less convenient for comparing when the values of x are small (Szmidt and Kacprzyk [193]). This fact is illustrated in Figure 4.12 – the bigger n, the flatter the similarity measures $(\frac{1-x^n}{1+x^n})$ for smaller values of x. It means that formal satisfaction of some mathematical assumptions is necessary but may be not sufficient for using a measure.

Another function which may be applied is the exponential function (cf. Pal and Pal [128])

$$g(x) = e^{-x} \quad (4.85)$$

giving for our function (4.73) (Szmidt and Kacprzyk [193])

$$Sim_4(X,F) = Sim(l^1_{IFS}(X,F), l^1_{IFS}(X,F^C)) =$$

$$= \frac{e^{-f(l^1_{IFS}(X,F), l^1_{IFS}(X,F^C))} - e^{-1}}{1 - e^{-1}} \quad (4.86)$$

It is obvious that one could continue generating more complicated functions $g(x)$ (being the decreasing functions of f) but it would not give any additional insight as far as similarity is concerned.

From the measures (4.80) – (4.86) intuitively acceptable results are obtained. Some examples, troublesome for other measures, are presented in Table 4.2. It

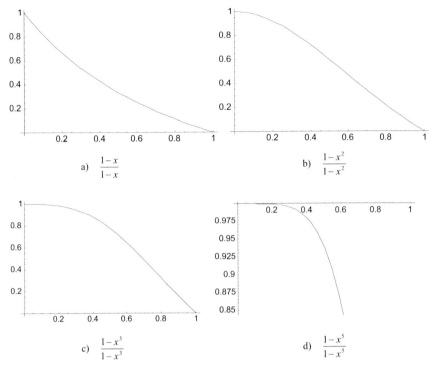

a) $\dfrac{1-x}{1-x}$ b) $\dfrac{1-x^2}{1-x^2}$ c) $\dfrac{1-x^3}{1-x^3}$ d) $\dfrac{1-x^5}{1-x^5}$

Fig. 4.12 The shapes of $\dfrac{1-x^n}{1+x^n}$

Table 4.2 Examples of results obtained from the similarity measures (4.80)–(4.86)

	1	2	3	4	5
$X = (\mu, \nu, \pi)$	$(0.3, 0.4, 0.3)$	$(0.4, 0.2, 0.4)$	$(0.4, 0.2, 0.4)$	$(0, 0, 0)$
$F = (\mu, \nu, \pi)$	$(0.4, 0.3, 0.3)$	$(0.5, 0.3, 0.2)$	$(0.5, 0.2, 0.3)$	$(0.5, 0.5, 0)$
Sim_1	0	0.6	0.75	0.5
Sim_2	0	0.43	0.6	0.33
Sim_3	0	0.72	0.88	0.6
Sim_4	0	0.48	0.65	0.38

is worth noticing that each measure assigns similarity equal 0 for an element $(0.3, 0.4, 0.3)$ and its complement $(0.4, 0.3, 0.3)$. In general, similarity measures (4.80) – (4.86) satisfy the following properties:

$$Sim_i(X, F) \in [0, 1], \tag{4.87}$$
$$Sim_i(X, X) = 1, \tag{4.88}$$
$$Sim_i(X, X^C) = 0, \tag{4.89}$$
$$Sim_i(X, F) = Sim_i(F, X), \tag{4.90}$$

for $i = 1, \ldots, 4$.

The similarity measures discussed in this section assess similarity of any two elements (X and F) belonging to an intuitionistic fuzzy set (or sets). The corresponding similarity measures for the intuitionistic fuzzy sets A and B, containing n elements each, are:

$$Sim_k(A,B) = \frac{1}{n}\sum_{i=1}^{n}Sim_k(l^1_{IFS}(X_i,F_i),l^1_{IFS}(X_i,F^C_i)) \tag{4.91}$$

for $k = 1,\ldots,4$.

Although in the formulas presented above we used the normalized Hamming distance, it is possible to replace it by other kinds of distances.

To be more specific, function $f(l^1_{IFS}(X,F),l^1_{IFS}(X,F^C))$ (4.73), making use of the Hamming distance in (4.80) – (4.86), can be replaced by the corresponding function making use of the Euclidean distance, i.e.:

$$f(q^1_{IFS}(X,F),q^1_{IFS}(X,F^C)) = \frac{q^1_{IFS}(X,F)}{q^1_{IFS}(X,F) + q^1_{IFS}(X,F^C)} \tag{4.92}$$

where $q^1_{IFS}(X,F)$ is given by (3.157). For example, the measure corresponding to similarity measure (4.80), in which (4.92) instead of (4.73) was applied, is:

$$Sim_1(q^1_{IFS}(X,F),q^1_{IFS}(X,F^C)) = 1 - f(q^1_{IFS}(X,F),q^1_{IFS}(X,F^C)) =$$

$$= 1 - \frac{q^1_{IFS}(X,F)}{q^1_{IFS}(X,F) + q^1_{IFS}(X,F^C)}. \tag{4.93}$$

The measure, corresponding to similarity measure (4.82), in which (4.92) instead of (4.73) was applied, is:

$$Sim_2(q^1_{IFS}(X,F),q^1_{IFS}(X,F^C)) = \frac{1 - f(q^1_{IFS}(X,F),q^1_{IFS}(X,F^C))}{1 + f(q^1_{IFS}(X,F),q^1_{IFS}(X,F^C))} \tag{4.94}$$

The measure, corresponding to similarity measure (4.84), in which (4.92) instead of (4.73) was applied, is:

$$Sim_3(q^1_{IFS}(X,F),q^1_{IFS}(X,F^C)) = \frac{1 - f(q^1_{IFS}(X,F),q^1_{IFS}(X,F^C))^2}{1 + f(q^1_{IFS}(X,F),q^1_{IFS}(X,F^C))^2} \tag{4.95}$$

The measure, corresponding to similarity measure (4.86), in which (4.92) instead of (4.73) was applied, is:

$$Sim_4(q^1_{IFS}(X,F),q^1_{IFS}(X,F^C)) = \frac{e^{-f(q^1_{IFS}(X,F),q^1_{IFS}(X,F^C))} - e^{-1}}{1 - e^{-1}} \tag{4.96}$$

In the similarity measures discussed here (Section 4.4), the problem of symmetry between the membership, non-membership and hesitation margin (cf. Szmidt and Kreinovich [228]) was partly removed by introducing into the definitions of the

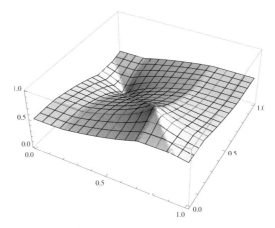

Fig. 4.13 Values obtained from measure (4.80) for any element from an intuitionistic fuzzy set and (0.7, 0.2, 0.1)

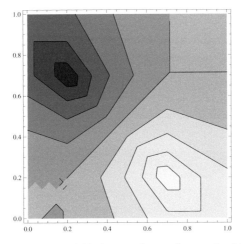

Fig. 4.14 Contourplot of measure (4.80) for any element from an intuitionistic fuzzy set and (0.7, 0.2, 0.1)

similarity measures not only the relation to an element we are interested in, but also that to its complement. In result, the measures discussed here better meet our expectations than the similarity measures being just concepts dual to distance (cf. Section 4.1). We avoid, for example, high values of similarity of an element and its complement. However, we still should use the similarity measures carefully.

In Figures 4.13–4.14 we have an illustration of the results generated by the similarity measure (4.80), while in Figures 4.15–4.16 the results given by the similarity measure (4.93) are presented. The results produced by both similarity measures (4.80) and (4.93) are analogous in the sense of pointing out some geometrical "shapes" but still the problem of symmetry concerning terms (describing an intuitionistic fuzzy set) in the formulas was not completely solved as quite different

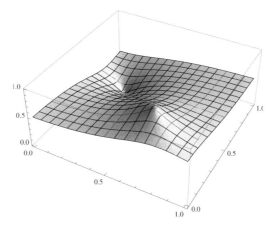

Fig. 4.15 Values obtained from (4.93) for any element from an intuitionistic fuzzy set and (0.7, 0.2, 0.1)

elements from the point of view of, for instance, decision making are "the same" in the sense of the values obtained from the proposed measures in respect to a chosen element. A simple "weighting" of the terms describing the elements does not solve the problem either. In Figures 4.17 and 4.18 we have the results from (4.93) with "weighted" membership values (in Figure 4.17 the membership value is two times more important, and in Figure 4.18 it is ten times more important than the non-membership and hesitation margin). The geometrical shapes implied by the weighted similarity measures change (cf. Figures 4.16, 4.17 and 4.18).

The question arises what should be done if we wish, e.g., to use similarity measure (4.93) and to differentiate between elements (0.3, 0, 0.7) and (0.5, 0.4, 0.1),

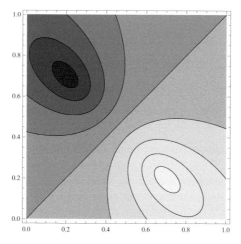

Fig. 4.16 Contourplot of (4.93) for any element from an intuitionistic fuzzy set and (0.7, 0.2, 0.1)

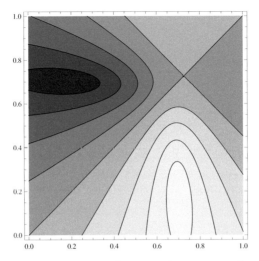

Fig. 4.17 Contourplot of (4.93) (with two times more important membership values) for any element from an intuitionistic fuzzy set and (0.7, 0.2, 0.1)

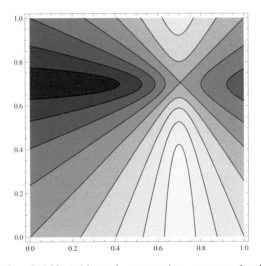

Fig. 4.18 Contourplot of (4.93) (with ten times more important membership values) for any element from an intuitionistic fuzzy set and (0.7, 0.2, 0.1)

which are obviously different from the point of view of decision making but both are similar to element (0.7, 0.2, 0.1) to the same extent equal to 0.6 (cf. Figures 4.15 and 4.16). Most important is that we should not conclude about similarity of (0.3, 0, 0.7) and (0.5, 0.4, 0.1) before calculating their direct similarity from (4.93), whereupon we obtain the value 0.51 (different from 0.6). This observation about examining similarity seems important when one tries to conclude about similarity of different elements on the basis of their direct distances to "the ideal" element (1,0,0).

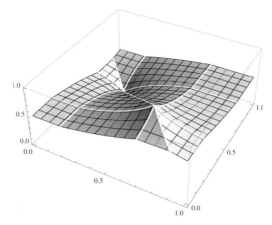

Fig. 4.19 Values of similarity (4.97) for any element from an intuitionistic fuzzy set and element (0.7, 0.2, 0.1)

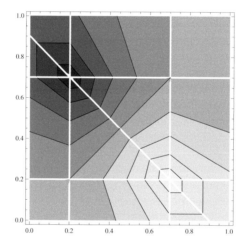

Fig. 4.20 Contourplot of measure (4.97) for any element from an intuitionistic fuzzy set and element (0.7, 0.2, 0.1)

Finally, we will introduce another definition of similarity using the Hausdorff distance (cf. Grünbaum [77]). We have shown (Section 3.3.2.1) that in the case of the Hausdorff distance between the intuitionistic fuzzy sets we should use a formula with all three terms describing the sets. If we apply distance (3.215) in the formulas (4.80) – (4.86) instead of (4.73), we obtain new similarity measures. For example, the counterpart of (4.80) with (3.215) replacing (4.73) is:

$$Sim(H_3(X,F), H_3(X,F^C)) = 1 - f(H_3(X,F), H_3(X,F^C)) =$$

$$= 1 - \frac{H_3(X,F)}{H_3(X,F) + H_3(X,F^C)}. \tag{4.97}$$

In Figures 4.19 and 4.20 we have exemplary results from (4.97) – the presence of the complement element and its influence on the results are visible.

Accounting for the complement elements in the similarity measures seems important in many tasks (for example, in image recognition, the most "dissimilar" image is a negative image which can be understood as an image consisting of the complement elements).

4.5 Correlation of Intuitionistic Fuzzy Sets

In Section 3.1 correlation coefficient was mentioned in the case of crisp sets as the standarized angular separation (3.13) resulting from centering the coordinates to the mean values. Here we will discuss the notion of correlation of intuitionistic fuzzy sets as introduced by Szmidt and Kacprzyk [211].

The correlation coefficient r (so called Pearson's coefficient) was proposed by Karl Pearson in 1895, and is one of the most used indices in statistics [140]. Correlation indicates how well two variables move together in a linear fashion, i.e., correlation reflects a linear relationship between two variables. Consequently, correlation coefficient became an important measure in data analysis and classification, in particular in decision making, medical diagnosis, predicting market behavior, pattern recognition, and other real world problems concerning political, legal, financial, economic, social, environmental, educational, artistic, etc. systems.

The concept has been extended further to fuzzy observations (cf. e.g., Chiang and Lin [50], Hong and Hwang [82], Liu and Kao [118]).

As relations between intuitionistic fuzzy sets (representing, e.g., preferences, attributes) seem to be of vital importance both in theory and practice, there are many papers discussing the correlation of the intuitionistic fuzzy sets: Bustince and Burillo [43], Gersternkorn and Mańko [73], Hong and Hwang [81], Hung [85], Hung and Wu [86], Zeng and Li [261]. Some of those papers evaluate only the strength of relationship (cf. Gersternkorn and Mańko [73], Hong and Hwang [81], Zeng and Li [261]). In other papers (cf. Hung [85], Hung and Wu [86]), a positive and negative type of relationship is reflected, but the third term describing an intuitionistic fuzzy set, which is important from the point of view of all distance, similarity, or entropy measures (cf. Szmidt and Kacprzyk, e.g., [165], [171], [188], [175], [192]), [193]) is not taken into account.

This section deals with a concept of correlation for data represented as intuitionistic fuzzy sets by adopting the concepts from statistics. We calculate it by showing both positive and negative relationships of the sets, and showing that it is important to take into account all three terms describing intuitionistic fuzzy sets.

The correlation coefficient (Pearson's r) between two variables is a measure of the linear relationship between them, equal to:

- 1 in the case of a positive (increasing) linear relationship,
- -1 in the case of a negative (decreasing) linear relationship,
- some value between -1 and 1 in all other cases.

The closer the coefficient is to either -1 or 1, the stronger the correlation between the variables.

Correlation between Crisp Sets

For a random sample of size n, i.e.: $(X_1, Y_1), (X_2, Y_2), \ldots, (X_n, Y_n)$ from a joint probability density function $f_{X,Y}(x,y)$, let $\overline{X}$ and $\overline{Y}$ be the sample means of variables X and Y, respectively, then the sample correlation coefficient $r(X,Y)$ is given as (e.g., [140]):

$$r(A,B) = \frac{\sum_{i=1}^{n} (x_i - \overline{X})(y_i - \overline{Y})}{(\sum_{i=1}^{n} (x_i - \overline{X})^2 \sum_{i=1}^{n} (y_i - \overline{Y})^2)^{0.5}} \tag{4.98}$$

where: $\overline{X} = \frac{1}{n} \sum_{i=1}^{n} x_i, \quad \overline{Y} = \frac{1}{n} \sum_{i=1}^{n} y_i.$

Correlation between Fuzzy Sets ([50])

For a random sample of size n, i.e. $x_1, x_2, \ldots, x_n \in X$ with a sequence of paired data $(\mu_A(x_1), \mu_B(x_1)), (\mu_A(x_2), \mu_B(x_2)), \ldots, (\mu_A(x_n), \mu_B(x_n))$ which correspond to the membership values of fuzzy sets A and B defined on X, the correlation coefficient $r_f(A,B)$ is defined as ([50]):

$$r_f(A,B) = \frac{\sum_{i=1}^{n} (\mu_A(x_i) - \overline{\mu_A})(\mu_B(x_i) - \overline{\mu_B})}{(\sum_{i=1}^{n} (\mu_A(x_i) - \overline{\mu_A})^2)^{0.5}(\sum_{i=1}^{n} (\mu_B(x_i) - \overline{\mu_B})^2)^{0.5}} \tag{4.99}$$

where: $\overline{\mu_A} = \frac{1}{n} \sum_{i=1}^{n} \mu_A(x_i), \quad \overline{\mu_B} = \frac{1}{n} \sum_{i=1}^{n} \mu_B(x_i).$

Correlation between Intuitionistic Fuzzy Sets (Szmidt and Kacprzyk [211])

We consider a correlation coefficient for two intuitionistic fuzzy sets, A and B, so that we could express not only a relative strength but also a positive or negative relationship between A and B. Next, we take into account all three terms describing an intuitionistic fuzzy set (membership, non-membership and the hesitation margins) because each of them influences the results.

Suppose that we have a random sample $x_1, x_2, \ldots, x_n \in X$ with a sequence of paired data $[(\mu_A(x_1), \nu_A(x_1), \pi_A(x_1)), (\mu_B(x_1), \nu_B(x_1), \pi_B(x_1))], [(\mu_A(x_2), \nu_A(x_2), \pi_A(x_2)), (\mu_B(x_2), \nu_B(x_2), \pi_B(x_2))], \ldots, [(\mu_A(x_n), \nu_A(x_n), \pi_A(x_n)), (\mu_B(x_n), \nu_B(x_n), \pi_B(x_n))]$ which correspond to the membership values, non-memberships values and hesitation margins of the intuitionistic fuzzy sets A and B defined on X, then the correlation coefficient $r_{A-IFS}(A,B)$ is given by Definition 4.2.

Definition 4.2. The correlation coefficient $r_{A-IFS}(A,B)$ between two intuitionistic fuzzy sets, A and B in X, is:

$$r_{A-IFS}(A,B) = \frac{1}{3}(r_1(A,B) + r_2(A,B) + r_3(A,B)) \tag{4.100}$$

where

$$r_1(A,B) = \frac{\sum\limits_{i=1}^{n}(\mu_A(x_i) - \overline{\mu_A})(\mu_B(x_i) - \overline{\mu_B})}{(\sum\limits_{i=1}^{n}(\mu_A(x_i) - \overline{\mu_A})^2)^{0.5}(\sum\limits_{i=1}^{n}(\mu_B(x_i) - \overline{\mu_B})^2)^{0.5}} \tag{4.101}$$

$$r_2(A,B) = \frac{\sum\limits_{i=1}^{n}(\nu_A(x_i) - \overline{\nu_A})(\nu_B(x_i) - \overline{\nu_B})}{(\sum\limits_{i=1}^{n}(\nu_A(x_i) - \overline{\nu_A})^2)^{0.5}(\sum\limits_{i=1}^{n}(\nu_B(x_i) - \overline{\nu_B})^2)^{0.5}} \tag{4.102}$$

$$r_3(A,B) = \frac{\sum\limits_{i=1}^{n}(\pi_A(x_i) - \overline{\pi_A})(\pi_B(x_i) - \overline{\pi_B})}{(\sum\limits_{i=1}^{n}(\pi_A(x_i) - \overline{\pi_A})^2)^{0.5}(\sum\limits_{i=1}^{n}(\pi_B(x_i) - \overline{\pi_B})^2)^{0.5}} \tag{4.103}$$

where: $\overline{\mu_A} = \frac{1}{n}\sum\limits_{i=1}^{n}\mu_A(x_i)$, $\overline{\mu_B} = \frac{1}{n}\sum\limits_{i=1}^{n}\mu_B(x_i)$, $\overline{\nu_A} = \frac{1}{n}\sum\limits_{i=1}^{n}\nu_A(x_i)$,

$\overline{\nu_B} = \frac{1}{n}\sum\limits_{i=1}^{n}\nu_B(x_i)$, $\overline{\pi_A} = \frac{1}{n}\sum\limits_{i=1}^{n}\pi_A(x_i)$, $\overline{\pi_B} = \frac{1}{n}\sum\limits_{i=1}^{n}\pi_B(x_i)$.

In the correlation coefficient (4.100) two factors play an important role:

- the amount of information expressed by the membership and non-membership degrees (4.101)–(4.102), and
- the reliability of information, expressed by the hesitation margins (4.103).

Remark: It should be emphasized that analogously as for the crisp and fuzzy data, $r_{A-IFS}(A,B)$ makes sense for the intuitionistic fuzzy variables whose values vary. If, for instance, the temperature is constant and the amount of ice cream sold does not change, then it is impossible to conclude about their relationship (as, from the mathematical point of view, we avoid zero in the denominator).

The correlation coefficient $r_{A-IFS}(A,B)$ (4.100) fulfills the following properties:

1. $r_{A-IFS}(A,B) = r_{A-IFS}(B,A)$

2. If $A = B$ then $r_{A-IFS}(A,B) = 1$

3. $|r_{A-IFS}(A,B)| \leq 1$

The above properties are not only fulfilled by the correlation coefficient $r_{A-IFS}(A,B)$ (4.100) but also by its every component (4.101)–(4.103).

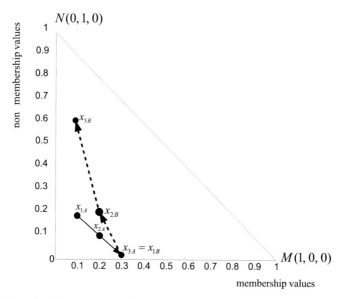

Fig. 4.21 Example 4.6: we can see from the data that there is no perfect negative linear relationship between elements from A and B

Remark: It is worth mentioning that $r_{A-IFS}(A,B) = 1$ occurs not only for $A = B$ but also in the cases of a perfect linear correlation of the data (cf. Example 4.7) (the same concerns each component (4.101)–(4.103)).

Now some simplified examples will be shown. The data set size is too small to look at them as at significant samples, but the purpose is just illustration.

Example 4.6. (Szmidt and Kacprzyk [211])
Let A and B be intuitionistic fuzzy sets in $X = \{x_1, x_2, x_3\}$:

$$A = \{(x_1, 0.1, 0.2, 0.7), (x_2, 0.2, 0.09, 0.71), (x_3, 0.3, 0.01, 0.69)\}$$

$$B = \{(x_1, 0.3, 0, 0.7), (x_2, 0.2, 0.2, 0.6), (x_3, 0.1, 0.6, 0.3)\}$$

Examining the above data in details, it is easy to notice that

- the membership values of the elements in A (i.e.: $0.1, 0.2, 0.3$) increase whereas the membership values of the elements in B (i.e.: $0.3, 0.2, 0.1$) decrease. In the result (4.101) we have $r_1(A, B) = -1$;
- the non-membership values of the elements in A (i.e.: $0.2, 0.09, 0.01$) decrease whereas the non-membership values of the elements in B (i.e.: $0.0, 0.2, 0.6$) increase. In the result (4.102) we have $r_2(A, B) \approx -0.96$.
- the hesitation margins of the elements in A (i.e.: $(0.7, 0.71, 0.0.69)$ and the hesitation margins of the elements in B (i.e.: $0.7, 0.6, 0.2$) give the result (4.103) $r_3(A, B) = 0.73$.

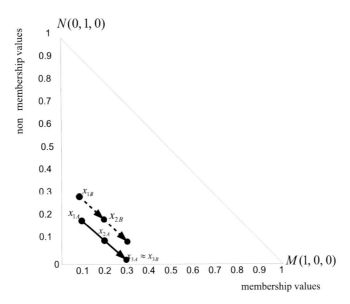

Fig. 4.22 Example 4.7: we can see from the data the perfect positive linear relationship among elements from A and B

Therefore, finally, from (4.100) we obtain $r_{A-IFS}(A,B) = \frac{1}{3}(-1 - 0.96 + 0.73) = -0.41$.

If we exclude from considerations the hesitation margins, and take into account two components (4.101) and (4.102) only, we obtain $r_{A-IFS}(A,B) = \frac{1}{2}(-1-0.96) = -0.98$ which means that there is a substantial negative linear relationship between A and B (which is difficult to agree with).

A geometrical interpretation (cf. Section 2.3.1) of the data from Example 4.6 is given in Figure 4.21.

It should be emphasized that, e.g., in decision making or other practical purposes it seems important to know the third component (4.103) of the correlation coefficient (4.100), concerning lack of knowledge represented by the variables considered. For example, if the data represent reactions of patients to a new medicine, it seems highly useful to examine carefully the component (4.103) of the correlation coefficient (4.100) as it may happen that a new medicine/treatment increases unforseen reactions. In such situations it may be important not only to assess all components separately but even to give them different weights in (4.100).

On the other hand, it is possible to give examples in which $r_3(A,B)$ does not influence the final result (the obtained value) of the correlation coefficient $r_{A-IFS}(A,B)$. But such situations are exceptions, not a rule.

Example 4.7. (Szmidt and Kacprzyk [211]) Let A and B be intuitionistic fuzzy sets in $X = \{x_1, x_2, x_3\}$:

Table 4.3 "Saturday Morning" data [138] in terms of intuitionistic fuzzy sets

No.	Attributes				Class
	Outlook	Temperature	Humidity	Windy	
1	$(0, 0.33, 0.67)$	$(0, 0.33, 0.67)$	$(0, 0.33, 0.67)$	$(0.2, 0, 0.8)$	N
2	$(0, 0.33, 0.67)$	$(0, 0.33, 0.67)$	$(0, 0.33, 0.67)$	$(0, 0.33, 0.67)$	N
3	$(1, 0, 0)$	$(0, 0.33, 0.67)$	$(0, 0.33, 0.67)$	$(0.2, 0, 0.8)$	P
4	$(0.2, 0.11, 0.69)$	$(0, 0, 1)$	$(0, 0.33, 0.67)$	$(0.2, 0, 0.8)$	P
5	$(0.2, 0.11, 0.69)$	$(0.4, 0.11, 0.49)$	$(0.6, 0, 0.4)$	$(0.2, 0, 0.8)$	P
6	$(0.2, 0.11, 0.69)$	$(0.4, 0.11, 0.49)$	$(0.6, 0, 0.4)$	$(0, 0.33, 0.67)$	N
7	$(1, 0, 0)$	$(0.4, 0.11, 0.49)$	$(0.6, 0, 0.4)$	$(0, 0.33, 0.67)$	P
8	$(0, 0.33, 0.67)$	$(0, 0, 1)$	$(0, 0.33, 0.67)$	$(0.2, 0, 0.8)$	N
9	$(0, 0.33, 0.67)$	$(0.4, 0.11, 0.49)$	$(0.6, 0, 0.4)$	$(0.2, 0, 0.8)$	P
10	$(0.2, 0.11, 0.69)$	$(0, 0, 1)$	$(0.6, 0, 0.4)$	$(0.2, 0, 0.8)$	P
11	$(0, 0.33, 0.67)$	$(0, 0, 1)$	$(0.6, 0, 0.4)$	$(0, 0.33, 0.67)$	P
12	$(1, 0, 0)$	$(0, 0, 1)$	$(0, 0.33, 0.67)$	$(0, 0.33, 0.67)$	P
13	$(1, 0, 0)$	$(0, 0.33, 0.67)$	$(0.6, 0, 0.4)$	$(0.2, 0, 0.8)$	P
14	$(0.2, 0.11, 0.69)$	$(0, 0, 1)$	$(0, 0.33, 0.67)$	$(0, 0.33, 0.67)$	N

$$A = \{(x_1, 0.1, 0.2, 0.7), (x_2, 0.2, 0.1, 0.7), (x_3, 0.29, 0.0, 0.71)\}$$

$$B = \{(x_1, 0.1, 0.3, 0.6), (x_2, 0.2, 0.2, 0.6), (x_3, 0.29, 0.1, 0.61)\}$$

It is easy to notice that

- the membership values of the elements in A (i.e.: $0.1, 0.2, 0.29$) increase, the membership values of the elements in B (i.e.: $0.1, 0.2, 0.29$) are the same as in A, so from (4.101) we have $r_1(A, B) = 1$.
- the non-membership values of the elements in A (i.e.: $0.2, 0.1, 0.$) decrease and the non-membership values of the elements in B (i.e.: $0.3, 0.2, 0.1$) decrease, and from (4.102) we have $r_2(A, B) = 1$.
- the hesitation margins of the elements in A are equal to $(0.7, 0.7, 0.71)$, and the hesitation margins of the elements in B are equal to $(0.6, 0.6, 0.61)$, so from (4.103) we have $r_3(A, B) = 1$.

Finally, from (4.100) we obtain $r_{A-IFS}(A, B) = \frac{1}{3}(1 + 1 + 1) = 1$. It is easy to notice that now the result is just the same whether we take into account $r_3(A, B)$ or not (i.e. when we consider (4.101) and (4.102) only, and divide their sum by 2). But, in general, $r_3(A, B)$ plays an important role in the correlation coefficient.

A geometric interpretation (cf. Section 2.3.1) of the data from Example 4.7 is shown in Figure 4.22. We can notice the perfect positive linear relationship between the elements from A and B (the perfect positive linear relationship of hesitation margins is expressed by the parallel lines formed by the elements from A and B – the two lines are also parallel to MN segment).

Now, following the results in (Szmidt and Kacprzyk [224]) we will verify on a more complicated example - the well known "Saturday Morning" [138] benchmark – if all the three parts of (4.100) play an important role. The data set is small

Table 4.4 Values of the correlation component (4.101) between each pair of attributes for the "Saturday Morning" data from [138]

Attr	Outlook	Temperature	Humidity	Windy
Outlook	1	0.01	0.03	-0.01
Temperature	0.01	1	0.63	-0.1
Humidity	0.03	0.63	1	0
Windy	-0.01	- 0.1	0	1

Table 4.5 Values of the correlation component (4.102) between each pair of attributes for the "Saturday Morning" data from [138]

Attr	Outlook	Temperature	Humidity	Windy
Outlook	1	0.01	0.12	-0.07
Temperature	0.01	1	0.12	-0.22
Humidity	0.12	0.12	1	0
Windy	-0.07	-0.22	0	1

Table 4.6 Values of the correlation component (4.103) between each pair of attributes for the "Saturday Morning" from [138]

Attr	Outlook	Temperature	Humidity	Windy
Outlook	1	0.15	-0.005	0.09
Temperature	0.15	1	0.45	-0.1
Humidity	-0.005	0.45	1	0
Windy	0.09	-0.1	0	1

and hence illustrative. Moreover, we know which interrelations to expect – three attributes are not strongly related (as each of them is important from the point of view of classification), the fourth one is not very important from the point of view of classification (more correlated with the others).

There are 14 examples in the "Saturday Morning" [138] data set, 4 nominal attributes, and the target attribute with two classes. The nominal attributes are: **outlook**, with values {sunny, overcast, rain}, **temperature**, with values {cold, mild, hot}, **humidity**, with values {high, normal}, and **windy**, with values {true, false}.

Making use of the idea presented in Section 2.5.2 (also in Szmidt and Baldwin [162], and in Szmidt and Kacprzyk [213]), we have obtained "Saturday Morning" data [138] description in terms of intuitionistic fuzzy sets (Table 4.3), i.e., we have expressed each attribute in terms of the membership values, non-membership values, and hesitation margin values. Next, we have calculated the three components of the correlation coefficient (4.100) for each pair of the attributes. The results are provided in Tables 4.4–4.6.

It is easy to notice that the correlation component (4.101) resulting from the correlation stemming from the membership values of attributes is significant only in one case (just as we have expected) – between Humidity and Temperature. The second component of (4.100), i.e., (4.102), in practically all cases produces values which

Table 4.7 Values of the correlation (4.100) between each pair of attributes for the "Saturday Morning" data from [138]

Attr	Outlook	Temperature	Humidity	Windy
Outlook	1	0.06	0.05	0.003
Temperature	0.06	1	0.4	-0.14
Humidity	0.05	0.4	1	0
Windy	0.003	-0.14	0	1

Table 4.8 Values of the first correlation component $r_1(A,B)$ (4.101) between each pair of attributes for Pima data

Attr	plasmgl	blopre	triceps	serins	bmi	dpf	age
pregn	0.15	0	-0.1	-0.1	-0.03	0.02	0.66
plasmgl	-	0.04	0.08	0.24	0.14	0.09	0.25
blopre	-	-	-0.05	-0.1	0.18	0.02	0.02
triceps	-	-	-	0.36	0.32	0.14	-0.1
serins	-	-	-	-	0.08	0.16	-0.05
bmi	-	-	-	-	-	0.1	-0.03
dpf	-	-	-	-	-	-	0.09

Table 4.9 Values of the second correlation component $r_2(A,B)$ (4.102) between each pair of attributes for the Pima Indians data

Attr	plasmgl	blopre	triceps	serins	bmi	dpf	age
pregn	-0.07	-0.13	-0.2	-0.14	-0.08	0.04	0.03
plasmgl	-	0.06	0.11	-0.08	0.12	-0.07	-0.02
blopre	-	-	0.04	0	0.15	0.03	0.02
triceps	-	-	-	0.42	0.19	0	0
serins	-	-	-	-	-0.12	0.03	-0.09
bmi	-	-	-	-	-	0.04	0.15
dpf	-	-	-	-	-	-	0

are not significant in terms of correlation. Next, the third component of (4.100), i.e., (4.103), confirms the conclusions we have drawn from (4.101). In other words, the values of the correlation expressed in terms of lack of knowledge (4.103) count, and should not be excluded from considerations when examining the correlation between the attributes.

We have obtained similar results for Pima Indians Diabetes Database [264].

Namely, we tried verify if the situation is similar (if all the three parts of $r_{A-IFS}(A,B)$ (4.100) count) for a well known benchmark example - the Pima Indian Diabetes Database [264]. The data set, known as Pima, contains 768 data examples in total, and 8 continuous attributes plus the target attribute with two classes. The continuous attributes are: number of times pregnant (pregn), plasma glucose concentration (plasmgl), diastolic blood pressure (blopre), triceps skin fold thickness (triceps), 2-hour serum insulin (serins), body mass index (bmi), diabetes pedigree function (dpf), age (age). The class distribution of the database is 500 data examples for class 1 and 268 data examples for class 2.

Table 4.10 Values of the third correlation component $r_3(A,B)$ (4.103) between each pair of attributes for the Pima data

Attr	plasmgl	blopre	triceps	serins	bmi	dpf	age
pregn	0.04	-0.08	-0.21	-0.13	-0.10	0.03	0.55
plasmgl	-	-0.02	0	0.02	0	-0.01	0.09
blopre	-	-	-0.06	-0.07	0.15	0.02	-0.06
triceps	-	-	-	0.44	0.12	0.08	-0.19
serins	-	-	-	-	-0.08	0.13	-0.12
bmi	-	-	-	-	-	0.03	-0.08
dpf	-	-	-	-	-	-	0.07

Table 4.11 Values of the total correlation $r_{A-IFS}(A,B)$ (4.100) between each pair of attributes for the Pima data

Attr	plasmgl	blopre	triceps	serins	bmi	dpf	age
pregn	0.04	-0.07	-0.17	-0.12	-0.07	0.03	0.41
plasmgl	-	0.03	0.06	0.06	0.09	0	0.11
blopre	-	-	-0.03	-0.06	0.16	0.02	-0.01
triceps	-	-	-	0.41	0.21	0.07	-0.09
serins	-	-	-	-	-0.04	0.11	-0.09
bmi	-	-	-	-	-	0.06	0.02
dpf	-	-	-	-	-	-	0.05

We used the algorithm based on the mass assignment theory proposed by Szmidt and Baldwin [162] to describe the data in terms of the intuitionistic fuzzy sets, i.e., to derive the parameters of an intuitionistic fuzzy set model which describes each attribute in terms of membership values, non-membership values, and hesitation margin values. Having description of the attributes in terms of intuitionistic fuzzy sets, we have calculated the three components of $r_{A-IFS}(A,B)$ (4.100) for each pair of the attributes. The results are given in Tables 4.8–4.10 (Szmidt et al. [226]).

It is easy to notice (Szmidt et al. [226]) that the attributes *pregn* and *age* are strongly correlated [0.66 for $r_1(A,B)$ (4.101) and 0.55 for $r_3(A,B)$ (4.103)]. Also the attributes *triceps* and *serins* are positively correlated [0.36 for $r_1(A,B)$ (4.101), 0.42 for $r_2(A,B)$ (4.102), 0.44 for $r_3(A,B)$ (4.103)]. The attribute *triceps* and *bmi* are also more significantly correlated [0.32 for $r_1(A,B)$ (4.101)] than the remaining pairs of attributes. We may notice again that the values $r_3(A,B)$ (4.103) are significant for the attributes mentioned. However, the significance of $r_3(A,B)$ (4.103) does not necessarily mean its substantial values - this depends on the values of $r_1(A,B)$ (4.101) and $r_2(A,B)$ (4.102). If both $r_1(A,B)$ (4.101) and $r_2(A,B)$ (4.102) are similar (and, e.g. "big", then "small" values of $r_3(A,B)$ (4.103) have influence on $r_{A-IFS}(A,B)$ (4.100) - see, e.g, *plasmgl* and *serins*, *plasmgl* and *age* - in effect of small values of both $r_2(A,B)$ (4.102) and $r_3(A,B)$ (4.103), the total correlation $r_{A-IFS}(A,B)$ (4.100) is small although in the studies of some real populations [78] *plasmgl* reflects risk for diabetes and is correlated with the mentioned attributes (which is reflected by $r_1(A,B)$ (4.101)). This fact speaks again for a careful insight into each component

Table 4.12 Values of the first correlation component $r_1(A,B)$ (4.101) between each pair of attributes for the Iris Setosa data

Attribute	sepal length	sepal width	petal length	petal width
sepal length	1	0.3	0.86	0.84
sepal width	-	1	0.6	0.6
petal length	-	-	1	0.99
petal width	-	-	-	1

Table 4.13 Values of the second correlation component $r_2(A,B)$ (4.102) between each pair of attributes for the Iris Setosa data

Attribute	sepal length	sepal width	petal length	petal width
sepal length	1	0.2	0.85	0.83
sepal width	-	1	0.51	0.5
petal length	-	-	1	0.99
petal width	-	-	-	1

Table 4.14 Values of the third correlation component $r_3(A,B)$ (4.103) between each pair of attributes for the Iris Setosa data

Attribute	sepal length	sepal width	petal length	petal width
sepal length	1	-0.14	0.67	0.61
sepal width	-	1	0.2	-0.14
petal length	-	-	1	0.6
petal width	-	-	-	1

Table 4.15 Values of the total correlation $r_{A-IFS}(A,B)$ (4.100) between each pair of attributes for the Iris Setosa data

Attribute	sepal length	sepal width	petal length	petal width
sepal length	1	0.13	0.79	0.76
sepal width	-	1	0.43	0.32
petal length	-	-	1	0.86
petal width	-	-	-	1

$r_1(A,B)$, $r_2(A,B)$, $r_3(A,B)$ (4.101)–(4.103). Especially that for diverse populations correlation coefficient between the pairs of the attributes vary [78]. Identification of the attributes having different association with incidence of diabetes reflects distinct metabolic processes about which important information may be lost easily when not examined in detail through $r_1(A,B)$, $r_2(A,B)$, and $r_3(A,B)$ (4.101)–(4.103).

We have also examined the correlation coefficient using Iris data [265] expressed in terms of the intuitionistic fuzzy sets (just the same as for the Pima data previously). Iris data consist of 3 classes with 50 instances each. Each class refers to a variety of the iris plant (Iris Setosa, Iris Versicolor, Iris Virginica). There are four attributes: sepal length, sepal width, petal length, petal width. Tables 4.12–4.24 show

Table 4.16 Values of the first correlation component $r_1(A,B)$ (4.101) between each pair of attributes for the Iris Versicolor data

Attribute	sepal length	sepal width	petal length	petal width
sepal length	1	0.35	0.81	0.77
sepal width	-	1	0.36	0.36
petal length	-	-	1	0.94
petal width	-	-	-	1

Table 4.17 Values of the second correlation component $r_2(A,B)$ (4.102) between each pair of attributes for the Iris Versicolor data

Attribute	sepal length	sepal width	petal length	petal width
sepal length	1	-0.09	0.82	0.72
sepal width	-	1	-0.07	-0.15
petal length	-	-	1	0.92
petal width	-	-	-	1

Table 4.18 Values of the third correlation component (4.103) between each pair of attributes for the Iris Versicolor data

Attribute	sepal length	sepal width	petal length	petal width
sepal length	1	0.2	0.8	0.61
sepal width	-	1	0.32	0.28
petal length	-	-	1	0.76
petal width	-	-	-	1

Table 4.19 Values of the total correlation $r_{A-IFS}(A,B)$ (4.100) between each pair of attributes for the Iris Versicolor data

Attribute	sepal length	sepal width	petal length	petal width
sepal length	1	0.16	0.8	0.7
sepal width	-	1	0.2	0.17
petal length	-	-	1	0.88
petal width	-	-	-	1

Table 4.20 Values of the first correlation component $r_1(A,B)$ (4.101) between each pair of attributes for the Iris Virginica data

Attribute	sepal length	sepal width	petal length	petal width
sepal length	1	0.25	0.6	0.51
sepal width	-	1	0.51	0.56
petal length	-	-	1	0.89
petal width	-	-	-	1

the results. We have examined the components $r_1(A,B)$, $r_2(A,B)$, $r_3(A,B)$ (4.101)–(4.103) of $r_{A-IFS}(A,B)$ (4.100) with respect to each class first. Results for Iris Setosa are given in Tables 4.12–4.15. It is easy to see that the petal length and petal width attributes are considerably correlated with one another and with other attributes es-

Table 4.21 Values of the second correlation component $r_2(A,B)$ (4.102) between each pair of attributes for the Iris Virginica data

Attribute	sepal length	sepal width	petal length	petal width
sepal length	1	0.28	0.36	0.32
sepal width	-	1	0.49	0.52
petal length	-	-	1	0.89
petal width	-	-	-	1

Table 4.22 Values of the third correlation component $r_3(A,B)$ (4.103) between each pair of attributes for the Iris Virginica data

Attribute	sepal length	sepal width	petal length	petal width
sepal length	1	0.04	0.54	0.33
sepal width	-	1	0.17	0.29
petal length	-	-	1	0.34
petal width	-	-	-	1

Table 4.23 Values of total correlation $r_{A-IFS}(A,B)$ (4.100) between each pair of attributes for the Iris Virginica data

Attribute	sepal length	sepal width	petal length	petal width
sepal length	1	0.19	0.5	0.38
sepal width	-	1	0.39	0.46
petal length	-	-	1	0.7
petal width	-	-	-	1

Table 4.24 Values of total correlation $r_{A-IFS}(A,B)$ (4.100) between each pair of attributes for all Iris data

Attribute	sepal length	sepal width	petal length	petal width
sepal length	1	0.16	0.7	0.62
sepal width	-	1	0.34	0.31
petal length	-	-	1	0.81
petal width	-	-	-	1

pecially with respect to $r_1(A,B)$, $r_2(A,B)$ (4.101)–(4.102) – Tables 4.12–4.13. The component $r_3(A,B)$ (4.103) – Table 4.14 influences a little the common result, i.e. $r_{A-IFS}(A,B)$ (4.100) – Table 4.15, but the trend remains the same.

Results for Iris Versicolor are provided in Tables 4.16–4.19. We can observe the same trend (as for Iris Setosa) but with lower correlation of petal length and petal width with regard to sepal width – Tables 4.17–4.18 (the components $r_2(A,B)$ (4.102) and $r_3(A,B)$ (4.103), respectively). The same is reflected in Table 4.19 (in respect with (4.100)).

For Iris Virginica – Tables 4.20–4.23, petal length and petal width are again considerably correlated with each other (especially with respect to component $r_1(A,B)$

(4.101) – Table 4.20). As for the other classes, we may observe again low correlation between sepal length and sepal width.

The correlation among the attributes for all three classes of the Iris data are summarized in Table 4.24. Although the trend of correlation between the attributes is still preserved, the previous detailed results give more and better insight into the data.

So, to sum up, we have discussed a correlation coefficient between the intuitionistic fuzzy sets. The coefficient proposed, like Pearson's coefficient between crisp sets, measures the strength of relationship between the intuitionistic fuzzy sets, and shows if the sets are positively or negatively correlated. All three terms describing the intuitionistic fuzzy sets are taken into account (the membership, non-membership and hesitation margin). Each term plays an important role in data analysis and decision making, so that each of them should be reflected when assessing the correlation between the intuitionistic fuzzy sets.

4.6 Concluding Remarks

At the beginning of this chapter we considered the notion of similarity between the intuitionistic fuzzy sets as a dual concept to distance. Unfortunately, in the case of the intuitionistic fuzzy sets this well known concept does not meet our expectations.

Next, two groups of similarity measures between the intuitionistic fuzzy sets were recalled. First, we presented a whole array of similarity measures (known from the literature) for the intuitionistic fuzzy sets, viewed in terms of single intervals. Second, we considered measures being straightforward generalizations of those well known for the fuzzy sets. However, both approaches do not meet our expectations, and both give counter-intuitive results.

Then, we reconsidered our concept of similarity measure between the intuitionistic fuzzy sets accounting for all three terms describing an intuitionistic fuzzy set (membership, non-membership and hesitation margin), which is different from viewing an intuitionistic fuzzy set as a single interval. We also took into account the complements of the elements compared. We have applied this measure of similarity to assess the extent of agreement in a group of experts giving their opinions expressed by intuitionistic fuzzy preference relations. We emphasized the intuitive appeal of the measure.

Further, we considered several modified similarity measures but still following the philosophy of employing all three terms describing the intuitionistic fuzzy sets, and making use of the complement elements. This may be viewed as an attempt of using all kinds and fine shades of information available. These last measures are the most promising, because, first of all, they help to avoid some strongly counter-intuitive results. This is crucial for both theory and applications. However, we have also pointed out the situations in which the measures should not be applied.

Finally, we presented an extended analysis of a Person's like correlation coefficient between the intuitionistic fuzzy sets. The coefficient proposed, like Pearson's coefficient between crisp sets, measures the strength of relationship between the intuitionistic fuzzy sets, and shows whether the sets are positively or negatively correlated. All three terms describing the intuitionistic fuzzy sets are taken into account (the membership value, non-membership value and hesitation margin). Each term plays an important role in data analysis and decision making, so that each should be reflected in the assessment of the correlation between the intuitionistic fuzzy sets.

Chapter 5
Summary and Conclusions

The intuitionistic fuzzy sets are a generalization of fuzzy sets with an additional degree of freedom, as compared to fuzzy sets, which are fully described by the degree of membership. In the definition of an intuitionistic fuzzy set a degree of non-membership is added, and the value of membership plus the value of non-membership for an element does not necessarily make one. Some psychological experiments demonstrate that in many judgments of human beings such a phenomenon happens. The additional degree of freedom means inherent possibility to model and process more adequately and more human consistently the imprecise information, and makes the intuitionistic fuzzy sets a useful tool in decision making.

The ability of expressing imprecise information leads to construction of more reliable models. The use of these models is connected with processing of imprecise information via different measures. The measures of distance and similarity are the basic and extremely important tools in processing of information.

In this book we dealt with measures of distance and similarity for the intuitionistic fuzzy sets, having in mind not only their mathematical correctness but also their practical aspects.

From the point of view of practical applications, provision of an automatic method of deriving the intuitionistic fuzzy sets from data (from relative frequency distributions) seems useful, especially in the context of analyzing information contained in big data bases. The approach has been shown to be useful in the context of benchmark data sets.

Two kinds of intuitionistic fuzzy set representations were considered. First, only two terms, namely membership values and non-membership values were taken into account (the two term representation). Next, all three terms, namely, membership values, non-membership values, and hesitation margin values were accounted for (the three term representation). We have considered these two representations from both geometrical and analytical points of view. The three term representation seems to be more justified and intuitively appealing from the practical point of view (this fact having its roots in some analytical and geometrical aspects).

Definite problems have been shown concerning the Hausdorff distance, in which Hamming metric was applied while using the two term intuitionistic fuzzy set

E. Szmidt, *Distances and Similarities in Intuitionistic Fuzzy Sets,*
Studies in Fuzziness and Soft Computing 307,
DOI: 10.1007/978-3-319-01640-5_5, © Springer International Publishing Switzerland 2014

representation. It has been also demonstrated that the method of calculating the Hausdorff distances in the same way, which is correct for the interval-valued fuzzy sets, does not work for the intuitionistic fuzzy sets.

The three term representation of the intuitionistic fuzzy sets has proved its usefulness when applied (as a component) in a measure of ranking of the intuitionistic fuzzy alternatives.

It has been also demonstrated that distance alone cannot be treated as a reliable concept dual to similarity in the case of the intuitionistic fuzzy sets.

Also the similarity measures between the intuitionistic fuzzy sets formulated as straightforward generalizations of those well known for fuzzy sets, do not meet our expectations. The situation is the same for another group of similarity measures for the intuitionistic fuzzy sets, viewed in terms of single intervals (which means using two terms only in the intuitionistic fuzzy set description).

On the other hand, the concept of the similarity measures between the intuitionistic fuzzy sets, accounting for all the three terms (membership, non-membership and hesitation margin), and taking into account the complements of the elements compared, avoids some counter-intuitive results, and meet better our expectations.

The same conclusion concerns Pearson's like correlation coefficient between the intuitionistic fuzzy sets – taking into account all of the three terms (the membership value, the non-membership value and the hesitation margin) is justified.

Summing up: intuitionistic fuzzy sets seem to be a comprehensive tool for handling many aspects of imprecise information. By taking into account the hesitation margin values besides the membership and non-membership values in construction of measures of distance and similarity, we can ensure better behavior and higher intuitive appeal of the measures considered.

References

1. Aichholzer, O., Alt, H., Rote, G.: Matching shapes with a reference point. Int. J. of Computational Geometry and Applications 7, 349–363 (1997)
2. Apolloni, B., Pedrycz, W., Bassis, S., Malchiodi, D.: The Puzzle of Granular Computing. SCI. Springer (2008)
3. Atallah, M.J.: A linear time algorithm for the Hausdorff distance between convex polygons. Information Processing Letters 17, 207–209 (1983)
4. Atanassov, K.: Intuitionistic Fuzzy Sets. VII ITKR Session. Sofia (Deposed in Centr. Sci.-Techn. Library of Bulg. Acad. of Sci. (1697/84) (1983) (in Bulgarian)
5. Atanassov, K.: Intuitionistic fuzzy relations. In: Third Int. Symp. "Automation and Sci. Instrumentation, Proc. part II, Varna, pp. 56–57 (1984)
6. Atanassov, K.: Intuitionistic Fuzzy Sets. Fuzzy Sets and Systems 20, 87–96 (1986)
7. Atanassov, K.: More on intuitionistic fuzzy sets. Fuzzy Sets and Systems 33, 37–46 (1989)
8. Atanassov, K.: Norms and metrics over intuitionistic fuzzy sets. Busefal 55, 11–20 (1993)
9. Atanassov, K.: New operations defined over the intuitionistic fuzzy sets. Fuzzy Sets and Systems 61, 137–142 (1994)
10. Atanassov, K.: Operators over interval valued intuitionistic fuzzy sets. Fuzzy Sets and Systems 64, 159–174 (1994)
11. Atanassov, K.: Norms and metrics over intuitionistic fuzzy logics. Busefal 59, 49–58 (1994b)
12. Atanassov, K.: On the geometric interpretations of the intuitionistic fuzzy logical objects. Part I. Busefal 60, 48–50 (1994)
13. Atanassov, K.: On the geometric interpretations of the intuitionistic fuzzy logical objects. Part II. Busefal 60, 51–54 (1994)
14. Atanassov, K.: On the geometric interpretations of the intuitionistic fuzzy logical objects. Part III. Busefal 60, 55–59 (1994)
15. Atanassov, K.: Intuitionistic Fuzzy Sets: Theory and Applications. Springer (1999)
16. Atanassov, K.: On intuitionistic fuzzy implication $\rightarrow^{\varepsilon}$ and intuitionistic fuzzy negation $\neg^{\varepsilon}$. Issues in Intuitionistic Fuzzy Sets and Generalized Nets 6, 6–19 (2008)
17. Atanassov, K.: Intuitionistic fuzzy implication $\rightarrow^{\varepsilon,\eta}$ and intuitionistic fuzzy negation $\neg^{\varepsilon,\eta}$. Developments in Fuzzy Sets, Intuitionistic Fuzzy Sets, Generalized Nets and Related Topics 1, 1–10 (2008)
18. Atanassov, K.: On the intuitionistic fuzzy implications and negations. Part 1. In: Cornelis, C., et al. (eds.) 35 Years of Fuzzy Set Theory - Celebratory Volume Dedicated to the Retirement of Etienne E. Kerre, pp. 19–38. Springer, Berlin (2010)
19. Atanassov, K., Dimitrov, D.: On the negations over intuitionistic fuzzy sets. Part 1. Annual of "Informatics" Section Union of Scientists in Bulgaria 1, 49–58 (2008)
20. Atanassov, K., Dimitrov, D.: Intuitionistic fuzzy implications and axioms for implications. Notes on Intuitionistic Fuzzy Sets 16(1), 10–20 (2010), http://ifigenia.org/wiki/issue:nifs/16/1/10-20
21. Atanassov, K.: Cantor's norms for intuitionistic fuzzy sets. Issues in Intuitionistic Fuzzy Sets and Generalized Nets 8, 36–39 (2010)
22. Atanassov, K.: On Intuitionistic Fuzzy Sets Theory. Springer (2012)
23. Atanassov, K., Burillo, P., Bustince, H.: On the intuitionistic fuzzy relations. Notes on Intuitionistic Fuzzy Sets 1(2), 87–92 (1995)
24. Atanassov, K., Gargov, G.: Interval-valued intuitionistic fuzzy sets. Fuzzy sets and Systems 31(3), 343–349 (1989)

25. Atanassov, K., Tasseva, V., Szmidt, E., Kacprzyk, J.: On the geometrical interpretations of the intuitionistic fuzzy sets. In: Atanassov, K., Kacprzyk, J., Krawczak, M., Szmidt, E. (eds.) Issues in the Representation and Processing of Uncertain and Imprecise Information. Fuzzy Sets, Intuitionistic Fuzzy Sets, Generalized Nets, and Related Topics, pp. 11–24. EXIT, Warsaw (2005)
26. Atanassova, L.: Remarks on the cardinality of the intuitionistic fuzzy sets. Fuzzy Sets and Systems 75(3), 399–400 (1995)
27. Atanassova, L., Atanassov, K.: An example for a "genuine" intuitionistic fuzzy set. In: Third Int. Symp. "Automation and Scientific Instrumentation", Proc. part II, Varna, pp. 58–60 (1984)
28. Baldwin, J.F.: Combining Evidences for Evidential Reasoning. International Journal of Intelligent Systems 6, 569–616 (1991)
29. Baldwin, J.F.: A Calculus for mass Assignments in Evidential Reasoning. In: Fedrizzi, M., Kacprzyk, J., Yager, R.R. (eds.) Advances in the Dempster-Shafer Theory of Evidence, pp. 513–531. John Wiley (1992a)
30. Baldwin, J.F.: The Management of Fuzzy and Probabilistic Uncertainties for Knowledge Based Systems. In: Shapiro, S.A. (ed.) Encyclopaedia of AI, 2nd edn., pp. 528–537. John Wiley (1992b)
31. Baldwin, J.F.: Mass assignments and fuzzy sets for fuzzy databases. In: Yager, R. (ed.) Advances in the Dempster-Shafer Theory of Evidence, pp. 577–594. John Wiley (1994)
32. Baldwin, J.F., Martin, T.P.: FRIL as an Implementation Language for Fuzzy Information Systems. In: IPMU 1996, Granada, pp. 289–294 (1996)
33. Baldwin, J.F., Pilsworth, B.W.: Semantic Unification with Fuzzy Concepts in Fril. In: IPMU 1990, Paris (1990)
34. Baldwin, J.F., Coyne, M.R., Martin, T.P.: Intelligent Reasoning Using General Knowledge to Update Specific Information: A Database Approach. Journal of Intelligent Information Systems 4, 281–304 (1995a)
35. Baldwin, J.F., Lawry, J., Martin, T.P.: A Mass Assignment Theory of the Probability of Fuzzy Events. ITRC Report 229, University of Bristol, UK (1995b)
36. Baldwin, J.F., Lawry, J., Martin, T.P.: Mass assignment based induction on decision trees of words. In: Proc. IPMU 1998, pp. 524–531 (1998c)
37. Baldwin, J.F., Lawry, J., Martin, T.P.: The Application of generalized Fuzzy Rules to Machine Learning and Automated Knowledge Discovery. Internationa Journal of Uncertainty, Fuzzyness and Knowledge-Based Systems 6(5), 459–487 (1998)
38. Baldwin, J.F., Martin, T.P., Pilsworth, B.W.: FRIL – Fuzzy and Evidential Reasoning in Artificial Intelligence. John Wiley (1995)
39. Bhattacharya, A.: On a measure of divergence of two multinomial populations. Sankhya 7, 401–406 (1946)
40. Bray, J.R., Curtis, J.T.: An ordination of the upland forest communities of Southern Wisconsin. Ecological Monographies 27, 325–349 (1957)
41. Bronshtein, I.N., Semendyayev, K.A., Musiol, G., Muehlig, H.: Handbook of Mathematics, 5th edn. Springer (2007)
42. Bujnowski, P.: Application of intuitionistic fuzzy sets for constructing decision trees for classification tasks. Ph.D. dissertation, SRI PAS, Warsaw (2013) (in Polish)
43. Bustince, H., Burillo, P.: Correlation of interval-valued intuitionistic fuzzy sets. Fuzzy Sets and Systems 74, 237–244 (1995)
44. Bustince, H., Burillo, P.: Vague sets are intuitionistic fuzzy sets. Fuzzy Sets and Systems 67, 403–405 (1996)
45. Bustince, H., Mohedano, V., Barrenechea, E., Pagola, M.: An algorithm for calculating the threshold of an image representing uncertainty through A-IFSs. In: IPMU 2006, pp. 2383–2390 (2006)

46. Bustince, H., Mohedano, V., Barrenechea, E., Pagola, M.: Image thresholding using intuitionistic fuzzy sets. In: Atanassov, K., Kacprzyk, J., Krawczak, M., Szmidt, E. (eds.) Issues in the Representation and Processing of Uncertain and Imprecise Information. Fuzzy Sets, Intuitionistic Fuzzy Sets, Generalized Nets, and Related Topics. EXIT, Warsaw (2005)

47. Bustince, H., Barrenechea, E., Pagola, M.: Image thresholding using restricted equivalence functions and maximizing the measures of similarity. Fuzzy Sets and Systems 158, 496–516 (2007)

48. Bustince, H., Barrenechea, E., Pagola, M.: Relationship between restricted dissimilarity functions, restricted equivalence functions and normal en-functions: Image thresholding invariant. Patter Recognition Letters 29, 525–536 (2008)

49. Carroll, J.D., Wish, M.: Multidimensional perceptual models and measurement methods. In: Carterette, E.C., Friedman, M.P. (eds.) Handbook of Perception. Academic Press, New York (1974)

50. Chiang, D.-A., Lin, N.P.: Correlation of fuzzy sets. Fuzzy Sets and Systems 102, 221–226 (1999)

51. Chen, S.M.: Measures of similarity between vague sets. Fuzzy Sets and Systems 74(2), 217–223 (1995)

52. Chen, S.M.: Similarity measures between vague sets and between elements. IEEE Trans. Syst. Mn Cybernet. 27(1), 153–158 (1997)

53. Chen, S.M., Tan, J.M.: Handling multi-criteria fuzzy decision-making problems based on vague-set theory. Fuzzy Sets and Systems 67(2), 163–172 (1994)

54. Clarke, K.R., Somerfield, P.J., Chapman, M.G.: On resemblance measures for ecological studies, including taxonomic dissimilarities and a zero-adjusted Bray-Curtis coefficient for denuded assemblages. Journal of Experimental Marine Biology and Ecology 330, 55–80 (2006)

55. Cross, V., Sudkamp, T.: Similarity and Compatibility in Fuzzy Set Theory. Physica-Verlag (2002)

56. Delgado, M., Moral, S.: On the concept of possibility-probability consistency. Fuzzy Sets and Systems 21, 311–318 (1987)

57. Dempster, A.P.: Upper and lower probabilities induced by a multivalued mapping. Annals of Mathematical Statistics 38(2), 325–339 (1967a)

58. Dempster, A.P.: Upper and lower probability inferences based on a sample from a finite univariate populattion. Biometrica 54(3), 515–528 (1967b)

59. Dempster, A.P.: A generalization of Bayesian inference. J. of the Royal Statistical Society, Series B 30, 205–247 (1968)

60. Li, D.-F.: Multiattribute decision making models and methods using intuitionistic fuzzy sets. Journal of Computer and System Sciences 70, 73–85 (2005)

61. Dubois, D., Prade, H.: Fuzzy Sets and Systems: Theory and Applications. Academic Press, Incorporated (1980)

62. Dubois, D., Prade, H.: On several representations of an uncertain body of evidence. In: Gupta, M.M., Sanchez, E. (eds.) Fuzzy Information and Decision Processes, pp. 167–181. North-Holland (1982)

63. Dubois, D., Prade, H.: Unfair coins and necessity measures: towards a possibilistic interpretation of histograms. Fuzzy Sets and Systems 10, 15–20 (1983)

64. Dubois, D., Prade, H.: The three semantics of fuzzy sets. Fuzzy Sets and Systems 90, 141–150 (1997)

65. Dubois, D., Prade, H.: Interval-valued fuzzy sets, possibility theory and imprecise probability. In: EUSFLAT-LFA 2005, pp. 314–319 (2005)

66. Dubois, D., Prade, H.: Fundamentals of fuzzy sets. Kluwer Academic Publishers, Boston (2000)

67. Dubois, D.: On degrees of truth, partial ignorance and contradiction. In: Magdalena, L., Ojeda-Aciego, M., Verdegay, J.M. (eds.) Proc. IPMU 2008, pp. 31–38 (2008)

68. Dubois, D., Foulloy, L., Mauris, G., Prade, H.: Probability-possibility transformations, triangular fuzzy sets, and probabilistic inequalities. Reliable Computing 10, 273–297 (2004)

69. Emran, S.M., Ye, N.: Robustness of Canberra metric in computer intrusion detection. In: Proceedings of the 2001 IEEE, Workshop on Information Assurance and Security, United States Military Academy, West Point, New York (2001)

70. Fan, J., Xie, W.: Distance measure and induced fuzzy entropy. Fuzzy Sets and Systems 104, 305–314 (1999)

71. Field, J.G., Clarke, K.R., Warwick, R.M.: A practical strategy for analysing multi-species distribution patterns. Marine Ecology Progress Series 8, 37–52 (1982)

72. Gau, W.L., Buehrer, D.J.: Vague sets. IEEE Trans. Systems Man Cybernet 23, 610–614 (1993)

73. Gersternkorn, T., Manko, J.: Correlation of intuitionistic fuzzy sets. Fuzzy Sets and Systems 44, 39–43 (1991)

74. Goguen, J.A.: L-fuzzy sets. Journal of Mathematical Analysis and Applications 18(1), 145–174 (1967)

75. Goguen, J.A.: The logic of inexact concepts. Synthese 19(3/4), 325–373 (1969)

76. Gomide, F., Pedrycz, W.: An Introduction to Fuzzy Sets: Analysis and Design. MIT Press, Cambridge (1998)

77. Grünmaum, B.: Convex Polytopes. Wiley Interscience, New York (1967)

78. Hanson, L.R., Imperatore, G., Bennett, P.H., Knowler, W.C.: Components of the "Metabolic Syndrome" and Incidence of Type 2 Diabetes. Diabetes 51, 3120–3127 (2002)

79. Higashi, M., Klir, G.: On measuresof fuzziness and fuzzy complements. International Journal General Systems 8, 169–180 (1982)

80. Hong, D.H., Choi, C.H.: Multicriteria fuzzy decision making problems based on vague set theory. Fuzzy Sets and Systems 114, 103–113 (2000)

81. Hong, D.H., Hwang, S.Y.: Correlation of intuitionistic fuzzy sets in probability spaces. Fuzzy Sets and Systems 75, 77–81 (1995)

82. Hong, D.H., Hwang, S.Y.: A note on the correlation of fuzzy numbers. Fuzzy Sets and Systems 79, 401–402 (1996)

83. Hong, D.H., Kim, C.: A note on similarity measures between vague sets and between elements. Inform Science 115, 83–96 (1999)

84. Hubálek, Z.: Coefficients of association and similarity, based on binary (presence-absence) data: An evaluation. Biological Reviews 57, 669–689 (1982)

85. Hung, W.L.: Using statistical viewpoint in developing correlation of intuitionistic fuzzy sets. Int. Journal of Uncertainty, Fuzziness and Knowledge-Based systems 9(4), 509–516 (2001)

86. Hung, W.L., Wu, J.W.: Correlation of intuitionistic fuzzy sets by centroid method. Information Sciences 144, 219–225 (2002)

87. Hung, W.-L., Yang, M.-S.: Similarity measures of intuitionistic fuzzy sets based on Hausdorff distance. Pattern Recognition Letters 25, 1603–1611 (2004)

88. Hung, W.L., Yang, M.S.: On similarity measures between intuitionistic fuzzy sets. International Journal of Intelligent Systems 23(3), 364–383 (2008)

89. Huttenlocher, D., Klanderman, G., Rucklidge, W.: Comparing images using the Hausdorff distance. IEEE Trans. on Pattern Analysis and Machine Intelligence 15(9), 850–863 (1993)

90. Huttenlocher, D., Rucklidge, W.: A multi–resolution technique for computing images using the Hausdorff distance. In: Proc. Computer Vision and Pattern Recognition, New York, pp. 705–708 (1993)

91. Kacprzyk, J.: Group decision making with a fuzzy linguistic majority. Fuzzy Sets and Systems 18, 105–118 (1986)

92. Kacprzyk, J., Fedrizzi, M.: 'Soft' consensus measures for monitoring real consensus reaching processes under fuzzy preferences. Control and Cybernetics 15, 309–323 (1986)

93. Kacprzyk, J., Fedrizzi, M.: A 'soft' measure of consensus in the setting of partial (fuzzy) preferences. European Journal of Operational Research 34, 315–325 (1988)

94. Kacprzyk, J., Fedrizzi, M.: A 'human-consistent' degree of consensus based on fuzzy logic with linguistic quantifiers. Mathematical Social Sciences 18, 275–290 (1989)

95. Kacprzyk, J., Fedrizzi, M., Nurmi, H.: Group decision making and consensus under fuzzy preferences and fuzzy majority. Fuzzy Sets and Systems 49, 21–32 (1992)

96. Kacprzyk, J., Fredizzi, M., Nurmi, H.: Fuzzy logic with linguistic quantifiers in group decision making. In: Yager, R., Zadeh, L. (eds.) An Introduction to Fuzzy Logic Applications in Intelligent Systems, pp. 263–280. Kluwer Academic Publishers, Norwell (1992)

97. Kacprzyk, J.: Multistage Fuzzy Control. Wiley, Chichester (1997)

98. Kahneman, D.: Maps of bounded rationality: a perspective on intuitive judgment and choice. Nobel Prize Lecture (2002)

99. Kaufmann, A.: Introduction to the theory of fuzzy sets. Academic Press, New York (1975)

100. Kelley, J.: General topology. D. van Nostrand Co., Toronto (1957)

101. Kendler, K.S., Josef Parnas, J.: Philosophical Issues in Psychiatry: Explanation, Phenomenology, and Nosology. Johns Hopkins University Press (2008)

102. Krebs, C.J.: Ecological Methodology. Harper-Collins, New-York (1989)

103. Lance, G.N., Williams, W.T.: Mixed-data classificatory programs I. Agglomerative Systems. Australian Computer Journal 1, 15–20 (1967)

104. Klir, G.J.: Where do we stand on measures of uncertainty, ambiguity, fuzziness, and the like? Fuzzy Sets and Systems 24, 141–160 (1987)

105. Klir, G.J.: Facets of Systems Science. Plenum Press, New York (1991)

106. Klir, G.J.: Uncertainty and Information. Foundations of Generalized Information Theory. John Wiley and Sons, Inc., New Jersey (2006)

107. Klir, G.J., Folger, T.A.: Fuzzy Sets, Uncertainty and Information. Prentice Hall, Englewood Cliffs (1988)

108. Klir, G.J., Wierman, M.J.: Uncertainty-Based Information. Elements of Generalized Information Theory. Physica-Verlag, Heidelberg (1998)

109. Klir, G.J., Yuan, B.: Fuzzy Sets and Fuzzy Logic. Theory and Applications. Prentice Hall PTR, New York (1995)

110. Lee, S.H., Pedrycz, W., Sohn, G.: Design of Similarity and Dissimilarity Measures for Fuzzy Sets on the Basis of Distance Measure. International Journal of Fuzzy Systems 11(2), 67–71 (2009)

111. Li, D.F., Cheng, C.T.: New similarity measures of intuitionistic fuzzy sets and application to pattern recognitions. Pattern Recognition Letters 23, 221–225 (2002)

112. Li, F., Lu, A., Cai, L.: Methods of multi-criteria fuzzy decision making base on vague sets. J. of Huazhong Univ. of Science and Technology 29(7), 1–3 (2001)

113. Li, Y., Olson, D.L., Qin, Z.: Similarity measures between intuitionistic fuzzy (vague) sets: A comparative analysis. Pattern Recognition Letters 28, 278–285 (2007)

114. Li, F., Rao, Y.: Weighted methods of multi-criteria fuzzy decision making based on vague sets. Computer Science 28(7), 60–65 (2001)

115. Li, Y., Zhongxian, C., Degin, Y.: Similarity measures between vague sets and vague entropy. J. Computer Sci. 29(12), 129–132 (2002)

116. Liang, Z., Shi, P.: Similarity measures on intuitionistic fuzzy sets. Pattern Recognition Lett. 24, 2687–2693 (2003)

117. Liu, X.: Entropy, distance mmeasure of fuzzy sets and their relations. Fuzzy Sets and Systems 52, 305–318 (1992)

118. Liu, S.-T., Kao, C.: Fuzzy measures for correlation coefficient of fuzzy numbers. Fuzzy Sets and Systems 128, 267–275 (2002)

119. Liu, H.-W., Wang, G.-J.: Multi-criteria decision making methods based on intuitionistic fuzzy sets. European Journal of Operational Research 179, 220 233 (2007)

120. Loewer, B., Laddaga, R.: Destroying the consensus. In: Loewer, B. (ed.) Special Issue on Consensus. Synthese, vol. 62, pp. 79–96 (1985)

121. Mahalanobis, P.C.: On the generalised distance in statistics. Proceedings of the National Institute of Sciences of India 2(1), 49–55 (1936)

122. McLachlan, G.J.: Discriminant Analysis and Statistical Pattern Recognition. Wiley Interscience (1992)

123. Mitchell, H.B.: On the Dengfeng-Chuntian similarity measure and its application to pattern recognition. Pattern Recognition Lett. 24, 3101–3104 (2003)

124. Montero, J., Gómez, D., Bustince, H.: On the relevance of some families of fuzzy sets. Fuzzy Sets and Systems 158, 2429–2442 (2007)

125. Moore, R.E.: Interval Analysis. Prentice-Hall, New York (1966)

126. Narukawa, Y., Torra, V.: Non-monotonic fuzzy measures and intuitionistic fuzzy sets. In: Torra, V., Narukawa, Y., Valls, A., Domingo-Ferrer, J. (eds.) MDAI 2006. LNCS (LNAI), vol. 3885, pp. 150–160. Springer, Heidelberg (2006)

127. Olson, C., Huttenlocher, D.: Automatic target recognition by matching oriented edge pixels. IEEE Trans. on Image Processing 6(1), 103–113 (1997)

128. Pal, N.R., Pal, S.K.: Entropy: a new definition and its applications. IEEE Trans. on Systems, Man, and Cybernetics 21(5), 1260–1270 (1991)

129. Pappis, C.P., Karacapilidis, N.: A comparative assessment of measures of similarity of fuzzy values. Fuzzy Sets and Systems 56, 171–174 (1993)

130. Paternain, D., Jurio, A., Bustince, H., Beliakov, G.: Image magnification using Atanassovs intuitionistic fuzzy sets

131. Jurio, A., Pagola, M., Mesiar, R., Beliakov, G., Bustince, H.: Image Magnification Using Interval Information. IEEE Transactiona on Image Processing 20(11), 3112–3123 (2011)

132. Pawlak, Z.: Rough sets. International Journal of Parallel Programming 11(5), 341–356 (1982)

133. Pedrycz, W.: Fuzzy Control and Fuzzy Systems, 2nd extended edn. Research Studies Press/John Wiley, Taunton (1993)

134. Pedrycz, W.: Fuzzy Sets Engineering. CRC Press, Boca Raton (1995)

135. Pedrycz, W., Gomide, F.: An Introduction to Fuzzy Sets. Analysis and Design. A Bradford Book. The MIT Press, Cambridge (1998)

136. Peitgen, H.O., Jürgens, H., Saupe, D.: Introduction to Fractals and Chaos. Springer, New York (1992)

137. Preparata, F.P., Shamos, M.I.: Computational Geometry. An Introduction. Springer, New York (1985)

138. Quinlan, J.R.: Induction of decision trees. Machine Learning 1, 81–106 (1986)

139. Ralescu, D.A.: Cardinality, quantifiers, and the aggregation of fuzzy criteria. Fuzzy Sets and Systems 69, 355–365 (1995)
140. Rodgers, J.L., Alan Nicewander, W.: Thirteen Ways to Look at the Correlation Coefficient. The American Statistician 42(1), 59–66 (1988)
141. Rote, G.: Computing the minimum Hausdorff distance between two point sets on a line under translation. Information Processing Letters 38, 123–127 (1991)
142. Rucklidge, W.J.: Lower bounds for the complexity of Hausdorff distance. Tech. report TR 94-1441, Dept. of computer science, Cornell University, NY. A similar title appeared in Proc. 5th Canad. Conf. on Comp. Geom (CCCG 1993), Waterloo, CA, pp. 145–150 (1995)
143. Rucklidge, W.J.: Efficient computation of the minimum Hausdorff distance for visual recognition. Ph.D. thesis, Dept. of computer science, Cornell University, NY (1995)
144. Rucklidge, W.J.: Locating objects using the Hausdorff distance. In: Proc. of 5th Int. Conf. on Computer Vision (ICCV 1995), Cambridge, MA, pp. 457–464 (1995)
145. Rucklidge, W.J.: Efficient Visual Recognition Using the Hausdorff Distance. LNCS, vol. 1173. Springer, Heidelberg (1996)
146. Rucklidge, W.J.: Efficiently locating objects using the Hausdorff distance. Int. Journal of Computer Vision 24(3), 251–270 (1997)
147. Rutkowski, L.: Computational Intelligence: Methods and Techniques. Springer, Heidelberg (2008)
148. Salton, G., McGill, M.J.: Introduction to Modern Information Retrieval. McGraw-Hill Book Company, New York (1983)
149. Schwartz, L.: Analyse Mathematique, Hermann, Paris (1967)
150. Shafer, G.: A Mathematical Theory of Evidence. Princeton Univ. Press (1976)
151. Shackle, G.L.: Decision, Order and Time in Human Affairs. Cambridge University Press, Edinburgh (1961)
152. Shepard, R.N.: Representation of structure in similarity data: Problems and prospects. Psychomelrika 39, 373–421 (1974)
153. Smets, P.: Constructing the pignistic probability function in a context of uncertainty. In: Henrion, M., Schachter, R., Kanal, L., Lemmer, J. (eds.) Uncertainty in Artificial Intelligence, vol. 5, pp. 29–39. North-Holland, Amsterdam (1990)
154. Sudkamp, T.: Similarity, interpolation, and fuzzy rule construction. Fuzzy Sets and Systems 58(1), 73–86 (1993)
155. Sugeno, M.: Theory of Fuzzy Integrals and its Applications. Tokyo Institute of Technology, Japan (1974)
156. Sugeno, M.: Fuzzy measures and fuzzy integrals: A survey. In: Gupta, M., Saridis, G., Gaines, B. (eds.) Fuzzy Automata and Decision Processes, pp. 89–102. North Holland, Amsterdam (1977)
157. Sutherland, S.: Irrationality. The Enemy Within. Penguin Books (1994)
158. Szmidt, E.: Applications of Intuitionistic Fuzzy Sets in Decision Making (D.Sc. dissertation), Technical University, Sofia (2000)
159. Szmidt, E., Baldwin, J.: New similarity measure for intuitionistic fuzzy set theory and mass assignment theory. Notes on Intuitionistic Fuzzy Sets 9(3), 60–76 (2003)
160. Szmidt, E., Baldwin, J.: Entropy for intuitionistic fuzzy set theory and mass assignment theory. Notes on Intuitionistic Fuzzy Sets 10(3), 15–28 (2004)
161. Szmidt, E., Baldwin, J.: Assigning the parameters for Intuitionistic Fuzzy Sets. Notes on Intuitionistic Fuzzy Sets 11(6), 1–12 (2005)
162. Szmidt, E., Baldwin, J.: Intuitionistic Fuzzy Set Functions, Mass Assignment Theory, Possibility Theory and Histograms. In: 2006 IEEE World Congress on Computational Intelligence, pp. 237–243 (2006)

163. Szmidt, E., Kacprzyk, J.: Intuitionistic fuzzy sets in group decision making. Notes on Intuitionistic Fuzzy Sets 2, 15–32 (1996a)
164. Szmidt, E., Kacprzyk, J.: Remarks on some applications of intuitionistic fuzzy sets in decision making. Notes on Intuitionistic Fuzzy Sets 2(3), 22–31 (1996c)
165. Szmidt, E., Kacprzyk, J.: On measuring distances between intuitionistic fuzzy sets. Notes on Intuitionistic Fuzzy Sets 3(4), 1–13 (1997)
166. Szmidt, E., Kacprzyk, J.: A Fuzzy Set Corresponding to an Intuitionistic Fuzzy Set. International Journal of Uncertainty, Fuzziness and Knowledge Based Systems 6(5), 427–435 (1998)
167. Szmidt, E., Kacprzyk, J.: Group Decision Making under Intuitionistic Fuzzy Preference Relations. In: Proc. IPMU 1998, Paris, La Sorbonne, pp. 172–178 (1998a)
168. Szmidt, E., Kacprzyk, J.: Applications of Intuitionistic Fuzzy Sets in Decision Making. In: EUSFLAT 1999, pp. 150–158 (1998b)
169. Szmidt, E., Kacprzyk, J.: Probability of Intuitionistic Fuzzy Events and their Applications in Decision Making. In: Proc. of EUSFLAT-ESTYLF, Palma de Mallorca, pp. 457–460 (1999)
170. Szmidt, E., Kacprzyk, J.: A Concept of a Probability of an Intuitionistic Fuzzy Event. In: Proc. of FUZZ-IEEE 1999 - 8th IEEE International Conference on Fuzzy Systems, Seoul, Korea, III, pp. 1346–1349 (1999b)
171. Szmidt, E., Kacprzyk, J.: Distances between intuitionistic fuzzy sets. Fuzzy Sets and Systems 114(3), 505–518 (2000)
172. Szmidt, E., Kacprzyk, J.: On Measures on Consensus Under Intuitionistic Fuzzy Relations. In: IPMU 2000, pp. 1454–1461 (2000)
173. Szmidt, E., Kacprzyk, J.: On Measures of Consensus Under Intuitionistic Fuzzy relations. In: Proc. IPMU 2000, Madrid, July 3-7, pp. 641–647 (2000b)
174. Szmidt, E., Kacprzyk, J.: Analysis of Consensus under Intuitionistic Fuzzy Preferences. In: Proc. Int. Conf. in Fuzzy Logic and Technology, pp. 79–82. De Montfort Univ, Leicester (2001)
175. Szmidt, E., Kacprzyk, J.: Entropy for intuitionistic fuzzy sets. Fuzzy Sets and Systems 118(3), 467–477 (2001)
176. Szmidt, E., Kacprzyk, J.: Distance from Consensus Under Intuitionistic Fuzzy Preferences. In: Proc. EUROFUSE Workshop on Preference Modeling and Applications, Granada, pp. 73–78 (2001)
177. Szmidt, E., Kacprzyk, J.: Analysis of Agreement in a Group of Experts via Distances Between Intuitionistic Fuzzy Preferences. In: Proc. 9th Int. Conference IPMU 2002, Annecy, France, pp. 1859–1865 (2002a)
178. Szmidt, E., Kacprzyk, J.: An Intuitionistic Fuzzy Set Based Approach to Intelligent Data Analysis (an application to medical diagnosis). In: Abraham, A., Jain, L., Kacprzyk, J. (eds.) Recent Advances in Intelligent Paradigms and Applications, pp. 57–70. Springer (2002c)
179. Szmidt, E., Kacprzyk, J.: Evaluation of Agreement in a Group of Experts via Distances Between Intuitionistic Fuzzy Sets. In: Proc. IS 2002 – Int. IEEE Symposium: Intelligent Systems, Varna, Bulgaria, IEEE Catalog Number 02EX499, pp. 166–170 (2002c)
180. Szmidt, E., Kacprzyk, J.: Similarity of intuitionistic fuzzy sets and the Jaccard coefficient. In: IPMU 2004, pp. 1405–1412 (2004)
181. Szmidt, E., Kacprzyk, J.: A Concept of Similarity for Intuitionistic Fuzzy Sets and its use in Group Decision Making. In: 2004 IEEE Conf. on Fuzzy Systems, Budapest, pp. 1129–1134 (2004)

182. Szmidt, E., Kacprzyk, J.: A New Concept of a Similarity Measure for Intuitionistic Fuzzy Sets and Its Use in Group Decision Making. In: Torra, V., Narukawa, Y., Miyamoto, S. (eds.) MDAI 2005. LNCS (LNAI), vol. 3558, pp. 272–282. Springer, Heidelberg (2005)

183. Szmidt, E., Kacprzyk, J.: A new Similarity Measure for Intuitionistic Fuzzy Sets and its use in Supporting a Medical Diagnosis. Notes on Intuitionistic Fuzzy Sets 11(4), 130–138 (2005)

184. Szmidt, E., Kacprzyk, J.: Distances Between Intuitionistic Fuzzy Sets and their Applications in Reasoning. SCI, vol. 2 (2005)

185. Szmidt, E., Kacprzyk, J.: New Measures of Entropy for Intuitionistic Fuzzy Sets. Notes on Intuitionistic Fuzzy Sets 11(2), 12–20 (2005)

186. Szmidt, E., Kacprzyk, J.: Similarity Measures for Intuitionistic Fuzzy Sets. In: Atanassov, K.T., Kacprzyk, J., Krawczak, M., Szmidt, E. (eds.) Issues in the Representation and Processing of Uncertain and Imprecise Information. Problems of Contemporary Science, pp. 355–372. EXIT, Warsaw (2005)

187. Szmidt, E., Kacprzyk, J.: A new measure of entropy and its connection with a similarity measure for intuitionistic fuzzy sets. In: Proceedings of the 4th Conference of the European Society for Fuzzy Logic and Technology (EUSFLAT 2005), Barcelona, pp. 1–6 (2005)

188. Szmidt, E., Kacprzyk, J.: Distances between intuitionistic fuzzy sets: straightforward approaches not work. In: 3rd International IEEE Conference Intelligent Systems, IS 2006, London, pp. 716–721 (May 2006)

189. Szmidt, E., Kacprzyk, J.: An Application of Intuitionistic Fuzzy Set Similarity Measures to a Multi-criteria Decision Making Problem. In: Rutkowski, L., Tadeusiewicz, R., Zadeh, L.A., Żurada, J.M. (eds.) ICAISC 2006. LNCS (LNAI), vol. 4029, pp. 314–323. Springer, Heidelberg (2006)

190. Szmidt, E., Kacprzyk, J.: A Model of Case Based Reasoning Using Intuitionistic Fuzzy Sets. In: 2006 IEEE World Congress on Computational Intelligence, pp. 8428–8435 (2006)

191. Szmidt, E., Kacprzyk, J.: Entropy and similarity for intuitionistic fuzzy sets. In: 11th Int. Conf. IPMU, Paris, pp. 2375–2382 (2006)

192. Szmidt, E., Kacprzyk, J.: Some Problems with Entropy Measures for the Atanassov Intuitionistic Fuzzy Sets. In: Masulli, F., Mitra, S., Pasi, G. (eds.) WILF 2007. LNCS (LNAI), vol. 4578, pp. 291–297. Springer, Heidelberg (2007)

193. Szmidt, E., Kacprzyk, J.: A New Similarity Measure for Intuitionistic Fuzzy Sets: Straightforward Approaches may not work. In: 2007 IEEE Conf. on Fuzzy Systems, pp. 481–486 (2007a)

194. Szmidt, E., Kacprzyk, J.: Classification with nominal data using intuitionistic fuzzy sets. In: Melin, P., Castillo, O., Aguilar, L.T., Kacprzyk, J., Pedrycz, W. (eds.) IFSA 2007. LNCS (LNAI), vol. 4529, pp. 76–85. Springer, Heidelberg (2007)

195. Szmidt, E., Kacprzyk, J.: Two and three parameter representation of intuitionistic fuzzy sets in the context of entropy and similarity. Notes on Intuitionistic Fuzzy Sets 13(2), 8–17 (2007)

196. Szmidt, E., Kacprzyk, J.: Dilemmas with Distances Between Intuitionistic Fuzzy Sets: Straightforward Approaches Not Work. SCI, vol. 109, pp. 415–430. Springer, Heidelberg (2008)

197. Szmidt, E., Kacprzyk, J.: Ranking alternatives expressed via intuitionistic fuzzy sets. In: 12th International Conference, IPMU 2008, pp. 1604–1611 (2008)

198. Szmidt, E., Kacprzyk, J.: A new approach to ranking alternatives expressed via intu-
     itionistic fuzzy sets. In: Ruan, D., et al. (eds.) Computational Intelligence in Decision
     and Control, pp. 265–270. World Scientific (2008)
199. Szmidt, E., Kacprzyk, J.: Intuitionistic fuzzy sets - a prospective tool for text catego-
     rization. In: Atanassov, K., Chountas, P., Kacprzyk, J., et al. (eds.) Developments in
     Fuzzy Sets, Intuitionistic Fuzzysets, Generalized Nets and Related Topics. Applica-
     tions, vol. II, pp. 281–300. Academc Pulisching House EXIT; Systems Research Insti-
     tute PAS, Warsaw (2008)
200. Szmidt, E., Kacprzyk, J.: Intuitionistic fuzzy sets as a promising tool for extended fuzzy
     decision making models. In: Bustince, H., Herrera, F., Montero, J. (eds.) Fuzzy Sets and
     Their Extensions: Representation, Aggregation and Models. STUDFUZZ, vol. 220, pp.
     330–355. Springer, Heidelberg (2008)
201. Szmidt, E., Kacprzyk, J.: Ranking intuitionistic fuzzy alternatives. Notes on Intuition-
     istic Fuzzy Sets 14(1), 48–56 (2008)
202. Szmidt, E., Kacprzyk, J.: Using intuitionistic fuzzy sets in text categorization. In:
     Rutkowski, L., Tadeusiewicz, R., Zadeh, L.A., Zurada, J.M. (eds.) ICAISC 2008. LNCS
     (LNAI), vol. 5097, pp. 351–362. Springer, Heidelberg (2008)
203. Szmidt, E., Kacprzyk, J.: Dealing with Typical Values by using Atanassov's Intuition-
     istic Fuzzy Sets. In: Proceedings of 2008 IEEE World Congress on Computational In-
     telligence, Hong Kong, June 1-6 (2008)
204. Szmidt, E., Kacprzyk, J.: On Some Typical Values for Atanassov's Intuitionistic Fuzzy
     Sets. In: Proc. of the 4th International IEEE Conference "Intelligent Systems", Varna,
     Bulgaria, vol. I, pp. 13-2-13-7 (2008)
205. Szmidt, E., Kacprzyk, J.: Amount of information and its reliability in the ranking
     of Atanassov's intuitionistic fuzzy alternatives. In: Rakus-Andersson, E., Yager, R.,
     Ichalkaranje, N., Jain, L.C. (eds.) Recent Advances in Decision Making. SCI, vol. 222,
     pp. 7–19. Springer, Heidelberg (2009)
206. Szmidt, E., Kacprzyk, J.: Ranking of Intuitionistic Fuzzy Alternatives in a Multi-criteria
     Decision Making Problem. In: Proceedings of the Conference, NAFIPS 2009, Cincin-
     nati, USA, June 14- 17, IEEE (2009) ISBN: 978-1-4244-4577-6
207. Szmidt, E., Kacprzyk, J.: Analysis of Similarity Measures for Atanassov's Intuitionistic
     Fuzzy Sets. In: Proceedings IFSA/EUSFLAT 2009, pp. 1416–1421 (2009)
208. Szmidt, E., Kacprzyk, J.: A method for ranking alternatives expressed via Atanassov's
     intuitionistic fuzzy sets. In: Atanassov, K.T., Hryniewicz, O., Kacprzyk, J., Krawczak,
     M., Nahorski, Z., Szmidt, E., Zadrony, S. (eds.) Advances in Fuzzy Sets, Intuitionistics
     Fuzzy Sets, Generalized Nets and Related Topics. Challenging Problems of Science -
     Computer Science, pp. 161–173. Academic Publishing House EXIT, Warsaw (2009)
209. Szmidt, E., Kacprzyk, J.: Some remarks on the Hausdorff distance between Atanassov's
     intuitionistic fuzzy sets. In: EUROFUSE WORKSHOP 2009. Preference Modelling
     and Decision Analysis, Pamplona (Spain), pp. 311–316. Public University of Navarra
     (2009)
210. Szmidt, E., Kacprzyk, J.: A note on the Hausdorff distance between Atanassov's intu-
     itionistic fuzzy sets. Notes on Intuitionistic Fuzzy Sets 15(1), 1–12 (2009)
211. Szmidt, E., Kacprzyk, J.: Correlation of intuitionistic fuzzy sets. In: Hüllermeier, E.,
     Kruse, R., Hoffmann, F. (eds.) IPMU 2010. LNCS, vol. 6178, pp. 169–177. Springer,
     Heidelberg (2010)
212. Szmidt, E., Kacprzyk, J.: The Spearman rank correlation coefficient between intuition-
     istic fuzzy sets. In: Proc. 2010 IEEE Int. Conf. on Intelligent Systems, IEEE'IS 2010,
     London, pp. 276–280 (2010)

213. Szmidt, E., Kacprzyk, J.: Dealing with typical values via Atanassov's intuitionistic fuzzy sets. International Journal of General Systems 39(5), 489–596 (2010)
214. Szmidt, E., Kacprzyk, J.: On an Enhanced Method for a More Meaningful Ranking of Intuitionistic Fuzzy Alternatives. In: Rutkowski, L., Scherer, R., Tadeusiewicz, R., Zadeh, L.A., Zurada, J.M. (eds.) ICAISC 2010, Part I. LNCS, vol. 6113, pp. 232–239. Springer, Heidelberg (2010)
215. Szmidt, E., Kacprzyk, J.: On the Hamming-metric based Hausdorff Distance for Intuitionistic Fuzzy Sets and Interval-valued Fuzzy Sets. In: Atanasov, K.T., et al. (eds.) Developments in Fuzzy Sets, Intuitionistics Fuzzy Sets, Generalized Nets and Related Topics, pp. 209–223. SRI PAS, Warszawa (2010)
216. Szmidt, E., Kacprzyk, J., Bujnowski, P.: On some measures of information and knowledge for intuitionistic fuzzy sets. Notes on Intuitionistic Fuzzy Sets 16(2), 1–11 (2010)
217. Szmidt, E., Kacprzyk, J., Bujnowski, P.: Information and Knowledge in the Context of Atanassov's Intuitionistic Fuzzy Set. In: Proc. of the 10th International Conference on Intelligent Systems Design and Applications, Cairo, Egypt, pp. 702–707 (2010)
218. Szmidt, E., Kacprzyk, J.: Intuitionistic fuzzy sets – Two and three term representations in the context of a Hausdorff distance. Acta Universitatis Matthiae Belii, Series Mathematics 19(19), 53–62 (2011), http://ACTAMTH.SAVBB.SK
219. Szmidt, E., Kacprzyk, J., Bujnowski, P.: Measuring the Amount of Knowledge for Atanassov's Intuitionistic Fuzzy Sets. In: Petrosino, A. (ed.) WILF 2011. LNCS (LNAI), vol. 6857, pp. 17–24. Springer, Heidelberg (2011)
220. Szmidt, E., Kacprzyk, J., Bujnowski, P.: Pearson's coefficient between intuitionistic fuzzy sets. Notes on Intuitionistic Fuzzy Sets 17(2), 25–34 (2011)
221. Szmidt, E., Kacprzyk, J., Bujnowski, P.: Pearson's Correlation Coefficient between Intuitionistic Fuzzy Sets: an Extended Theoretical and Numerical Analysis. In: Atanassov, K.T., et al. (eds.) Recent Advances in Fuzzy Sets, Intuitionistic Fuzzy Sets, Generalized Nets and Related Topics, pp. 223–236. SRI PAS, Warsaw (2011)
222. Szmidt, E., Kacprzyk, J.: The Kendall Rank Correlation between Intuitionistic Fuzzy Sets. In: Proc.: World Conference on Soft Computing, San Francisco, CA, USA, 23/05/2011–26/05/2011 (2011)
223. Szmidt, E., Kacprzyk, J.: The Spearman and Kendall rank correlation coefficients between intuitionistic fuzzy sets. In: Proc. 7th conf. European Society for Fuzzy Logic and Technology, pp. 521–528. Antantic Press, Aix-Les-Bains (2011)
224. Szmidt, E., Kacprzyk, J.: On an Enhanced Method for a More Meaningful Pearson's Correlation Coefficient between Intuitionistic Fuzzy Sets. In: Rutkowski, L., Korytkowski, M., Scherer, R., Tadeusiewicz, R., Zadeh, L.A., Zurada, J.M. (eds.) ICAISC 2012, Part I. LNCS, vol. 7267, pp. 334–341. Springer, Heidelberg (2012)
225. Szmidt, E., Kacprzyk, J.: A new approach to principal component analysis for intuitionistic fuzzy data sets. In: Greco, S., Bouchon-Meunier, B., Coletti, G., Fedrizzi, M., Matarazzo, B., Yager, R.R. (eds.) IPMU 2012, Part II. CCIS, vol. 298, pp. 529–538. Springer, Heidelberg (2012)
226. Szmidt, E., Kacprzyk, J., Bujnowski, P.: Correlation between Intuitionistic Fuzzy Sets: Some Conceptual and Numerical Extensions. In: WCCI 2012, IEEE World Congress on Computational Intelligence, Brisbane, Australia, pp. 480–486 (2012)
227. Szmidt, E., Kacprzyk, J., Bujnowski, P.: Advances in Principal Component Analysis for Intuitionistic Fuzzy Data Sets. In: 2012 IEEE 6th International Conference "Intelligent Systems", pp. 194–199 (2012)
228. Szmidt, E., Kreinovich, V.: Symmetry between true, false, and uncertain: An explanation. Notes on Intuitionistic Fuzzy Sets 15(4), 1–8 (2009)

229. Szmidt, E., Kukier, M.: Classification of Imbalanced and Overlapping Classes using Intuitionistic Fuzzy Sets. In: 3rd International IEEE Conference on Intelligent Systems, IS 2006, London, pp. 722–727 (2006)

230. Szmidt, E., Kukier, M.: A new approach to classification of imbalanced classes via Atanassov's intuitionistic fuzzy sets. In: Wang, H.-F. (ed.) Intelligent Data Analysis: Developing New Methodologies Through Pattern Discovery and Recovery, pp. 85–102. Idea Group (2008)

231. Szmidt, E., Kukier, M.: Atanassov's intuitionistic fuzzy sets in classification of imbalanced and overlapping classes. In: Chountas, P., Petrounias, I., Kacprzyk, J. (eds.) Intelligent Techniques and Tools for Novel System Architectures. SCI, vol. 109, pp. 455–471. Springer, Heidelberg (2008)

232. Szmidt, E., Kukier, M.: Intuitionistic fuzzy classifier a tool for recognizing imbalanced classes. In: Atanassov, K., et al. (eds.) New Developments in Fuzzy Sets, Intuitionistic Fuzzy Sets, Generalized Nets and Related Topics. Applications, vol. II, pp. 287–296. Systems Research Institute, Polish Academy of Sciences, Warsaw (2012)

233. Szmidt, E., Kacprzyk, J., Kukier, M.: Intuitionistic fuzzy classifier for imbalanced classes. In: Rutkowski, L., Korytkowski, M., Scherer, R., Tadeusiewicz, R., Zadeh, L.A., Zurada, J.M. (eds.) ICAISC 2013, Part I. LNCS (LNAI), vol. 7894, pp. 483–492. Springer, Heidelberg (2013)

234. Tan, C., Zhang, Q.: Fuzzy multiple attribute TOPSIS decision making method based on intuitionistic fuzzy set theory. In: Proc. IFSA 2005, pp. 1602–1605 (2005)

235. Tanev, D.: On an intuitionistic fuzzy norm. Notes on Intuitionistic Fuzzy Sets 1(1), 25–26 (1995)

236. Tasseva, V., Szmidt, E., Kacprzyk, J.: On one of the geometrical interpretations of the intuitionistic fuzzy sets. Notes on IFS 11(3), 21–27 (2005)

237. Tversky, A.: Features of similarity. Psychol. Rev. 84, 327–352 (1977)

238. Veltkamp, R.C., Hagedoorn, M.: Shape similarity measures, properties and constructions. In: Laurini, R. (ed.) VISUAL 2000. LNCS, vol. 1929, pp. 467–476. Springer, Heidelberg (2000)

239. Veltkamp, R.: Shape Matching: similarity measures and algorithms. In: Proc. Shape Modelling International, pp. 187–197. IEEE Press, Italy (2001)

240. Veltkamp, R.C.: Shape Matching: Similarity Measures and Algorithms. UU-CS-2001-3, 1–17 (2001)

241. Wang, W.J.: New similarity measures on fuzzy sets and on elements. Fuzzy Sets and Systems 85, 305–309 (1997)

242. Wang, X., Kerre, E.: Resonable properties for the ordering of fuzzy quantities (I). Fuzzy Sets and Systems 118, 375–385 (2001)

243. Wang, X., Kerre, E.: Resonable properties for the ordering of fuzzy quantities (II). Fuzzy Sets and Systems 118, 387–405 (2001)

244. Wang, X., De Baets, B., Kerre, E.: A comparative study of similarity measures. Fuzzy Sets and Systems 73(2), 259–268 (1995)

245. Willims, W.T., Dale, M.B.: Fundamental problems in numerical taxonomy. Advances in Botanical Research 2, 35–68 (1965)

246. Wolda, H.: Similarity indices, sample size and diversity. Oecologia 50, 296–302 (1981)

247. Wygralak, M.: Vaguely defined objects – Representations, Fuzzy Sets and Nonclassical Cardinality Theory. Kluwer, Dordrecht (1996)

248. Yager, R.R.: Level sets for membership evaluation of fuzzy subsets. Tech. Rep. RRY-79-14, Iona Colledge, New York (1979); Also in: Yager, R. (ed.) Fuzzy Set and Possibility Theory – Recent Developments. Pergamon Press, Oxford, 90–97 (1982)

249. Yager, R.R., Kacprzyk, J., Fedrizzi, M. (eds.): Advances in the Dempster-Shafer Theory of Evidence. Wiley, New York (1994)
250. Yager, R.R.: On measures of fuzziness and negation. Part I: Membership in the unit interval. International Journal General Systems 5, 221–229 (1997)
251. Yamada, K.: Probability–Possibility Transformation Based on Evidence Theory. In: Proc. IFSA–NAFIPS 2001, pp. 70–75 (2001)
252. Ye, J.: Cosine similarity measures for intuitionistic fuzzy sets and their applications. Mathematical and Computer Modelling 53, 91–97 (2011)
253. Xu, Z.: Intuitionistic preference relations and their application in group decision making. Information Sciences 177, 2363–2379 (2007)
254. Zadeh, L.A.: Fuzzy sets. Information and Control 8, 338–353 (1965)
255. Zadeh, L.A.: Similarity relations and fuzzy orderings. Information Sciences 3, 177–200 (1971)
256. Zadeh, L.A.: Fuzzy Sets as the Basis for a Theory of Possibility. Fuzzy Sets and Systems 1, 3–28 (1978)
257. Zadeh, L.A.: A computational approach to fuzzy quantifiers in natural languages. Computera and Mathematics with Applications 9, 149–184 (1983)
258. Zadeh, L.A.: Fuzzy probabilities. Information Processing and Management 20, 363–372 (1986)
259. Zadeh, L.A.: Fuzzy logic, neural networks, and soft computing. Communications of the ACM 3(3), 77–84 (1994a)
260. Zadeh, L.A.: Soft computing and fuzzy logic. IEEE Software 11(6), 48–56 (1994b)
261. Zeng, W., Li, H.: Correlation coefficient of intuitionistic fuzzy sets. Journal of Industrial Engineering International 3(5), 33–40 (2007)
262. Zimmermann, H.J.: Fuzzy Sets, Decision Making, and Expert Systems. Kluwer, Dordrecht (1987)
263. Zimmermann, H.J.: Fuzzy Set Theory and its Applications, 3rd edn. Kluver, Boston (1996)
264. http://archive.ics.uci.edu/ml/datasets/Diabetes
265. http://archive.ics.uci.edu/ml/datasets/Iris

Index

Printed in the United States
By Bookmasters